THE GREAT AMERICAN
Hall *of* Wonders

Smithsonian American Art Museum, Washington D.C.,
in association with D Giles Limited, London

Claire Perry

Art, Science, and Invention
in the Nineteenth Century

THE GREAT AMERICAN
Hall *of* Wonders

The Great American Hall of Wonders *is organized by
the Smithsonian American Art Museum in collaboration
with the United States Patent and Trademark Office.*

*Battelle has provided important leadership support
for the exhibition.*

*Generous contributions are also provided by:
Sheila Duignan and Mike Wilkins
The Raymond J. and Margaret Horowitz Endowment
Thelma and Melvin Lenkin
Betty and Whitney MacMillan
Jean Mahoney
Robin Martin*

THE GREAT AMERICAN
Hall *of* Wonders

Chief of Publications: Theresa J. Slowik
Editor: Jane McAllister
Production editor: Tiffany D. Farrell
Designer: Jessica L. Hawkins
Production Assistant: Megan C. Krefting
Photo Researcher: Jean Lawrence

 Smithsonian American Art Museum

The Smithsonian American Art Museum is home to one of the largest collections of American art in the world. Its holdings—more than 41,000 works—tell the story of America through the visual arts and represent the most inclusive collection of American art in any museum today. It is the nation's first federal art collection, predating the 1846 founding of the Smithsonian Institution. The museum celebrates the exceptional creativity of the nation's artists whose insights into history, society, and the individual reveal the essence of the American experience.

For more information or a catalogue of publications, write:
Office of Publications, Smithsonian American Art Museum,
MRC 970, PO Box 37012, Washington, DC 20013-7012

Visit the museum's Web site at AmericanArt.si.edu.

Library of Congress
Cataloging-in-Publication Data

Library of Congress Cataloging-in-Publication Data

Perry, Claire, 1954-
The great American hall of wonders : art, science, and invention in the nineteenth century / Claire Perry.
    p. cm.
"Published in conjunction with the exhibition *Great American Hall of Wonders*, on view at the Smithsonian American Art Museum in Washington, D.C., from July 15, 2011 through January 8, 2012."

Includes bibliographical references and index.
ISBN 978-1-904832-97-3 (cloth cover : alk. paper)
ISBN 978-0-9790678-9-1 (soft cover : alk. paper)

1. United States–Civilization–19th century–Exhibitions.
2. Creative ability–United States–History–19th century–Exhbitions.
3. Arts, American–19th century–Themes, motives–Exhibitions. I. Title.

E169.1.P443 2011
973.5--dc22

2011006182

Printed and bound in Belgium.

First published in 2011 by GILES, an imprint of D Giles Limited, in association with the Smithsonian American Art Museum.

GILES

D Giles Limited
4 Crescent Stables
139 Upper Richmond Road
London SW15 2TN
UK
www.gilesltd.com

Published in conjunction with the exhibition *Great American Hall of Wonders*, on view at the Smithsonian American Art Museum in Washington, D.C., from July 15, 2011 through January 8, 2012.

p. ii, Albert Bierstadt, *Giant Redwood Trees of California* (detail), about 1874. See fig. 111, p. 184.

p. iv, Unknown artist, *Man of Science* (detail), 1839. See fig. 3, p. 11.

p. xiv, Charles Willson Peale, *Exhumation of the Mastodon* (detail), about 1806–07. See fig. 1, p. 3.

p. 24, Francis William Edmonds, *Time to Go* (detail), 1857. See fig. 14, p. 30.

p. 44, Thomas Cole, *Distant View of Niagara Falls* (detail), 1830. See fig. 31, p. 58.

p. 76, William Michael Harnett, *The Faithful Colt* (detail), 1890. See fig. 67, p. 109.

p. 110, Ernest Griset, *The Far West.—Shooting Buffalo on the Line of the Kansas-Pacific Railroad.* (detail), 1871. See fig. 76, p. 122.

p. 132, Thomas Hill, *The Last Spike* (detail), 1875–80. See fig. 94, p. 156.

p. 160, Carleton Watkins, The Grizzly Giant, Mariposa Grove (detail), 1861. See fig. 108, p. 177.

p. 190, Winslow Homer, *Old Mill (The Morning Bell)* (detail), 1871. See fig. 18, pp. 38–39.

p. 206, Francis William Edmonds, *The Image Pedlar* (detail), about 1844. See fig. 49, p. 83.

p. 212, Thomas Cole, *The Clove, Catskills* (detail), about 1826. See fig. 103, p. 171.

p. 234, John Mix Stanley, *Buffalo Hunt on the Southwestern Prairies* (detail), 1845. See fig. 72, p. 118.

❧ Contents

❧ Foreword

Dr. Claire Perry's book, *The Great American Hall of Wonders*, seems at first to take a vagabond's route through nineteenth-century history. She opens with an almost surreal dinner party—hosted in 1802 by Philadelphian naturalist and painter Charles Willson Peale—at which the guests dined inside the ribcage of a mastodon, recently exhumed from native soil. From there, she creates a mix-and-match account of big trees and guns, clocks and Niagara Falls, buffalo and railroads, as if she were playing the old parlor game of Exquisite Corpse with a group of friends. "Backward reels the mind" as the reader tries to discern some linear thread in the tangle of astounding details and rich amazement of the story.

Along the way, it dawns that the tangle *is* the story. As Perry says in her preface, "Over the course of the century, the United States as we know it came into being, and from all indications the process was a seat-of-the-pants business." Like the emerging grassroots democracy that she describes, she takes a free-wheeling, individualistic approach to her story, and in doing so, she captures its essence.

Perry paints a picture of the early United States in psychic distress as the Founding Fathers died and left common citizens to carry forward our Great Experiment in democratic self-government. As our population swelled, diversity increased, and square mileage expanded, Americans struggled to unite the country around a shared understanding of democratic governance. There was, quite simply,

no model to follow, no book of instructions on how to mold a disorganized rabble into a citizenry.

The founding documents promised "life, liberty, and the pursuit of happiness," which implied limited government power, especially when further constrained by a Bill of Rights. "Pursuit of happiness" took an urgent turn as the natural riches of the continent were discovered and seized from tribal communities by new settlers. The Land of Opportunity—a mixture of easily available resources, ambition, and hard work, sometimes seasoned with greed—unleashed a torrent of can-do creativity that began to create a distinctive American way of looking at the world. Little by little, Americans came to define themselves by their spirit of invention and embrace of progress, including all of its unintended consequences.

Traveling in the United States from 1831 to 1832, Alexis de Tocqueville found Americans to be self-seekers who "form the habit of thinking of themselves in isolation and imagine that their whole destiny is in their own hands," leading to a disrespect of others and erosion of moral character. Yet he admired their tendency to form voluntary associations to solve every kind of problem, without appealing to government. Indeed, "Nothing strikes a European traveler in the United States more than the absence of what we would call government or administration."

At the conclusion of her mind-expanding mapping of the explosive intersection of nature and American invention, Perry quotes Ralph Waldo Emerson:

Alas for America as I so often say, the ungirt, the diffuse, the profuse, procumbent, one wide ground of juniper out of which no cedar, no oak will rear up a mast to the clouds! It all runs to leaves, to suckers, to tendrils, to miscellany.... America is formless.

Perry recaptures the daunting formlessness of our democracy without lapsing into Emerson's tone of lament. She takes a certain perverse joy in the nineteenth century's "ungirt suckers and tendrils," and discovers in them a distinctive genealogical lineage for our contemporary society to consider. While sizing up these American ancestors who were so busy with their pursuit of happiness, we are amused to find an occasional horse thief, and we're proud of the many over-achievers. Mostly we are fascinated to recognize our still-rather-formless selves in the legacy of democratic innovation. When we search for who we have become as a citizenry today, we find ourselves in this story.

My special thanks go to Dr. Perry for conceiving this exhibition and book, and for her splendid leadership and cooperation on all aspects of this project. She is an exceptional scholar devoted to original and large ideas in American art and history. I deeply appreciate also the generosity of our many donors to the project and the many helpful lenders to the exhibition. I also salute the hugely talented professionals at the Smithsonian American Art Museum who accomplished this complex project.

There could be no better collaborator for *The Great American Hall of Wonders* than the United States Patent and Trademark Office. The museum now occupies the noble historic building that was the Patent Office home during the Industrial Revolution, so our organizations share close historic bonds. The opportunity to work together on a project about the innovative character of our people is a joy.

Elizabeth Broun
The Margaret and Terry Stent Director
Smithsonian American Art Museum

❧ Preface

The United States began with an act of imagination. The Founding Fathers envisioned the nation as a Great Experiment for promoting human happiness, an ongoing trial of democratic liberties that would be sustained by the inquiring spirit of each new generation. At his presidential inauguration, George Washington spoke of an American revolution that never reached a conclusion. "The sacred fire of liberty, and the destiny of the republican model of government are . . . staked, on the experiment entrusted to the hands of the American people." After Thomas Jefferson and the last of the Founding Fathers died in the early nineteenth century, those in the new generation were both exhilarated and fearful when the delicate mechanisms of democracy, so carefully calibrated in the country's constitution, passed into their keeping. They began to call themselves "an inventive people" as they pondered the question: what, exactly, should their democratic nation be? Never ones to lollygag, they hastened to propose answers in their scientific and artistic descriptions of America's bounteous nature and in mechanical inventions aimed at improving their lives.[1]

The Great American Hall of Wonders: Art, Science, and Invention in the Nineteenth Century looks back to the nineteenth century's spirit of inquiry through science and technology, the arts, and materials of everyday life that defined the experimentations taking place in the vast laboratory of the United States. Prospecting through popular prints, paintings by eminent artists, illustrated science books, photographs, engineering drawings, and patent models, this book investigates six iconic themes that sparked creative brainstorming across a wide swath of nineteenth-century society. Three of those subjects—the buffalo, the giant sequoia, and Niagara Falls—represent imaginative domains in which the people of the United States formulated beliefs about nature and their nation's abundant natural resources. Within the creative domains of three technological objects—the clock, the gun, and the railroad—Americans began to place innovation at the center of their lives and to equate improvements in technology with the purposeful use of time. Railroad trestles, big trees, and Colt pistols inspired not only artists, scientists, and mechanics but also editors, farmwives, and members of Congress as they pieced together different meanings of "life, liberty and the pursuit of happiness."

This book focuses on the years from 1826 to 1876, a period when the people of the United States worried constantly about their ability to keep the Experiment going. By the end of the nineteenth century, the question seemed to be resolved in the public mind, as America was fast becoming the most powerful nation in the world. Between the death of Thomas Jefferson in 1826 and the Centennial Exposition of 1876, however, an overwhelming number of citizens expressed a strong and personal sense of responsibility for the nation's destiny. Americans believed that a wide range of knowledge was needed to undertake the work of democracy and saw useful material in an astonishing array of fields and skills. Everyday people devoted themselves to the study of geology, botany, paleontology, zoology, and statistics. Meteorology, the science of making predictions, became an American passion. Eager to increase the comforts of democratic life, hundreds of thousands of individuals applied to the Patent Office with the new devices they had invented in their spare time. Artists collaborated by reconnoitering the national territory and the diverse and far-flung communities that made up the body politic. Shaped by the aspirations of its stewards, the United States emerged as a great "hall of wonders," a showcase of natural abundance, freedom, and ingenuity.[2]

The Great American Hall of Wonders was inspired by the work of today's environmental historians, evolutionary biologists, art historians, economists, and scientific and technological innovators who trespass across disciplinary boundaries to ask questions about how we understand the world. Their insights are the foundation of this study.[3] The topic here is not science, art, or mechanical innovation per se, but rather what Americans of the nineteenth century believed about those endeavors and how they deployed them to direct their lives and the nation. Nineteenth-century Americans considered ingenuity to be their most important asset. The following chapters trace their creative meanderings through designs for locomotive spark arrestors and seed planters, paintings of frontier trappers, and covers for sheet music. The aim of the book has been to catch Americans making selections about what was possible and what

was not in a land of liberty. Their choices parceled out opportunities in varying measure to the nation's multifold communities and reconfigured its ecological systems in profound and irreversible ways. Over the course of the century the United States as we know it came into being, and from all indications the process was a seat-of-the-pants business.

Modern technologies have surpassed the nineteenth century's beginning efforts, and we well know that century's failures, from the extermination of the passenger pigeon to the devastation wreaked by the Civil War. But we are the heirs of that century's successful experiments as well as the ones that went terribly awry; we must look back to America's nineteenth-century stewards, for they have much to teach us. As we confront the complexities of our twenty-first century stewardship, knowing where we have come from may show us where we are headed. How do we sustain our nation's liberties, bountiful natural systems, and prosperity? If we accept the Founding Fathers' premise that the Great Experiment is refreshed each day by each American, we have work to do. Today's urgent social and environmental challenges call for a great national brainstorm, a collaborative imagining of enduring solutions. We can start by claiming our forebears' most important legacy—the belief in the transformative power of American inventiveness.

❧ Acknowledgments

The book and exhibition *The Great American Hall of Wonders* together represent the hard work of dozens of people. At the Smithsonian American Art Museum, Jane Paul, Claire Larkin, Eunice Kim, Theresa Slowik, Jessica Hawkins, Elaine Webster, Nona Martin, Laura Baptiste, and Susan Nichols devoted substantial time and know-how to the *Wonders* project. Its champion, Elizabeth Broun, promised to secure certain key works of art for the exhibition and worked tirelessly to make that happen. I am especially grateful to the museum's George Gurney and Rachel Allen, who contributed their tremendous expertise, wisdom, and good humor. At the National Archives, Jim Zeender offered his keen detective skills to locate important patent drawings. Bernard Barryte of the Cantor Arts Center at Stanford helped shape an early draft of the manuscript, and editor Jane McAllister worked long hours to polish its final form. The dedication and talent of Jeanie Lawrence allowed me to consider taking on the organizational challenges of the task in the first place.

It is possible to use "wonders" for the title of this book and exhibition only because nearly fifty museums, archives, libraries, and historical societies were willing to share their precious objects and works of art. For these institutions, the loans represent the sacrifice of a vacant place on a gallery wall or vitrine, as well as valuable staff time spent on loan administration. The generosity of the lenders is truly remarkable, and we acknowledge it with deep appreciation. We would especially like to recognize the outstanding contributions of the Pennsylvania Academy of the Fine Arts, Philadelphia; the National Gallery of Art, Washington, D.C.; the Maryland Historical Society, Baltimore; the Wadsworth Atheneum Museum of Art, Hartford, Connecticut; the National Museum of American History, Smithsonian Institution, Washington, D.C.; the Berkshire Museum, Pittsfield, Massachusetts; the Toledo Museum of Art, Ohio; the Iris and B. Gerald Cantor Center for Visual Arts at Stanford University, California; The Buffalo Bill Historical Center, Cody, Wyoming; Museum of the American West, Autry National Center of the American West, Los Angeles, California; and the Museum of the City of New York.

It has been a great privilege to participate in *The Great American Hall of Wonders*. I've spent the past several years immersed in its themes of imagination and creativity, and what could be luckier than that? My own source of inspiration is the love and affection of my family—Beau, Kristin, Byron, Somerset, Sebastian, Winslow, and all our wonderful extended clan.

To Oliver Crary, with gratitude

First let us suppose that we have before us a spacious building...in which are arranged specimens of all the various animals of this vast continent....We proceed step by step, to trace the beauties which we shall find that each possesses, in its relative situation to other beings.

Charles Willson Peale, *Discourse Introductory to a Course of Lectures on the Science of Nature* (1800)

The Hall of Wonders

N A FROSTY FEBRUARY NIGHT in 1802, CHARLES WILLSON Peale sat down to dinner with a group of friends and family inside the ribbed vault of his mastodon skeleton. The assembled well-wishers raised their glasses and sang "Yankee Doodle" to toast Peale's triumph in bringing the skeleton of the renowned mastodon to his museum in Philadelphia. As they engaged in lively conversation, the party would have seen the candles on the table cast shifting shadows of Pleistocene tusks and femurs on the walls. All knew that Peale's quest for what Thomas Jefferson had called the "great carnivorous elephant of the North" had consumed years of his life and now, inside the beast's ample belly, it was Peale's turn to tuck into a fine repast. The naturalist John

Godman described what it was like to first lay eyes on the magnificent monster. "The emotions experienced, when for the first time we behold the giant relics of this great animal, are those of unmingled awe," Godman noted. "We cannot avoid reflecting on the time when this huge frame was clothed with its peculiar instruments, and moved by appropriate muscles; when the mighty head dashed forth its torrents of blood through vessels of enormous caliber, and the mastodon strode along in supreme dominion over every other tenant of the wilderness."[4]

In the gaiety of the evening, the dinner guests may have exhorted Peale—museum founder, artist, scientist, inventor, and former saddle-maker—to describe the thrilling saga of his great find. Peale loved to tell the story and, to be sure that its drama would never be forgotten, would later spend two years completing a large canvas that he titled *Exhumation of the Mastodon* (fig. 1). The painting deploys more than seventy figures, a sky filled with darkening clouds, and an ominous lightning bolt to set the stage for an apocalyptic event. Encircling a watery pit, shovelers, bucket haulers, supervisors, and onlookers fill the lower half of the picture. Underneath the wheel of the enormous and ungainly pumping contraption that Peale himself had devised for the excavation, a worker arises, Lazarus-like, from the pit. He holds up a prodigious bone, and Peale, clutching a drawing of a similar ancient bone, gestures to him from the edge of the bog as if calling, yes, you have found it! An observer present at the site recounted the general hullabaloo that greeted the first discovery:

The unconscious woods echoed with repeated huzzahs, which could not have been more animated if every tree had participated in the joy. "Gracious God, what a jaw! How many animals have been crushed between it!" was the exclamation of all; a fresh supply of grog went round, and the hearty fellow, covered with mud, continued the search with increasing vigor.[5]

In Peale's painted recollection of that crowning moment, he imaginatively expanded the crowd of participants to include scientific colleagues and deceased loved ones whom he wished had been there. He depicted his wife, Hannah, and children, Rembrandt, Rubens, Raphaelle, Elizabeth, Sybilla, Titian Ramsey II, Linnaeus, and Franklin, as well as Jacob Masten, the owner of the New York farm on which the dig took place. Peale also featured his deceased second wife, Elizabeth

FIG. I

Charles Willson Peale
Exhumation of the Mastodon
about 1806–7, oil on canvas,
48 ¾ × 62 ½ in.
Maryland Historical Society,
Baltimore City Life Museum
Collection

De Peyster Peale, and the absent Alexander Wilson, then the nation's foremost ornithologist. Invoking the omnipotent authority of the artist, Peale altered time and space to share credit for the exhumation of the mastodon with people past and present. The sprinkling of his children's images across the canvas, conjuring with their singular names the accomplishments of artists, naturalists, and revolutionaries of times gone by, also proclaimed the generative nature of the great find and joined it to discoveries still to come.[6]

The skeleton from Masten's farm became the centerpiece of Peale's museum, a bracing introduction to its natural history specimens, works of art, and mechanical marvels. Dedicating his hall of wonders to the citizens of the United States, the artist orchestrated his displays to inspire them. Like most of the revolutionary generation, Peale believed that the longevity of the United States was not assured. On the contrary, it had been launched by the founders as a Great Experiment in democratic government that would survive only if the ideals of the Revolution were reinvigorated by the inquiring spirit of each subsequent generation. Peale's purpose with his museum was to kindle in his fellow citizens a deep regard for the beauty and usefulness of the boundless nature that was in their trust and to equip them thereby for the state-making tasks ahead.

In 1822, when Peale had reached the venerable age of eighty-one, the trustees of the museum commissioned him to paint a full-length self-portrait to honor his great contributions to science, art, education, and mechanical improvements in the United States. Peale took advantage of the commission to sum up his life's purpose and set forth a template of the work of citizenship. In the resulting *The Artist in His Museum* (fig. 2), Peale portrayed himself as a guide ready to embark on a tour of the treasures of the galleries and lifting a curtain to welcome a visitor. Behind him stretches the Long Room, a one-hundred-foot space filled with exhibits drawn from the vast continental laboratory where the nation's Great Experiment was unfolding. Within the Long Room, specimens were parsed into orderly categories that encouraged visitors to explore nature's meanings. Peale took great delight in the organization of his collections, which he believed was the key to making the material accessible to average citizens. Each bird, shell, insect, and rock had its proper place in the tidy rows of shelves and cabinets. They were laid out according to the Linnaean system of classes, orders, and genera that formed a "great chain of being" from lower to higher organisms, with humankind at the apex of its natural hierarchy.

Peale's repository looked back to the Old World's *Wunderkammern* and "cabinets of curiosity," chambers where princes assembled fine paintings, unicorn horns, gem-studded automata, and other treasures created by virtuoso artists and craftsmen and gathered on expeditions to the farthest reaches of the globe. Speaking of the collectors'

erudition, taste, and authority, the *Wunderkammer* elicited deference and put political rivals and visiting foreign dignitaries in their place.[7] Peale cribbed from these soul-stirring precedents but reconfigured their function to serve the purposes of democracy. The artist took great pains to ensure that citizens from all walks of life felt at ease among his marvels, and he even devised special lighting to allow the galleries to be open in the evening for the convenience of farmers, mechanics, and other laboring folk. Peale's egalitarian assembly of objects offered practical information for everyone. He explained that his museum provided the mechanic "accurate knowledge of many of the quantities of those materials with which his art is connected" and advised that "[for] the merchant, the study of nature is scarcely less interesting, whose traffic lies altogether in materials either raw from the stores of nature, or *wrought* by the hand of ingenious art."[8] Aspiring to convene the nation's multitudes around a shared esteem for nature, Peale proposed with his museum that a sense of wonder was the heart of the democratic project.

By 1820, Peale's galleries contained an impressive haul of more than ten thousand objects, including 121 fish, 148 snakes, 112 lizards, 40 tortoises and turtles, 1044 shells, 8,000 mineral specimens, 180 portraits, a wax figure of Meriwether Lewis, and a pair of revolving microscopes for viewing small specimens. *The Artist in His Museum* shows most clearly the museum's avian collection, numbering more than a thousand birds and still growing, of which the artist was most proud. On the left side of the scene, birds of varying sizes strike curious poses within a row of cubicles. The flock represented not only Peale's countless expeditions to remote areas and his competent marksmanship, but also his tireless experimentation with different preservation techniques and his close ties with friends, family members, and colleagues who thoughtfully brought back their finds to "good old Charlie." The most distinguished bird donor was George Washington, who gave Peale a pair of golden pheasants that had been a gift from the Marquis de Lafayette.

Using techniques he employed as a saddlemaker during his younger years, Peale learned how to stretch bird skins and feathers over wooden forms to shape dead creatures into lifelike attitudes. He then added glass eyes that he had molded himself, and placed his creations in cases that he and his children painted with naturalistic views of the bird's natural habitat. The lifeless turkey flopped upon the taxidermist's box at Peale's feet, a specimen that his son Titian brought back from his trip with the Long Expedition to Missouri, offers a hint of the pickling and posing taking place behind the scenes at the museum. On the threshold of the gallery, the forlorn gobbler awaits the reanimating art of his elderly keeper, the magic that is afoot behind the curtain.[9]

The feathered compartments in *The Artist in His Museum* are capped by a row of paintings, bust-length portraits of American worthies, framed in gold. The canvases represent the pantheon at the summit of the Linnean pecking order that governed the Long Room. Benjamin Franklin, George Washington, Thomas Jefferson, and other heads of state appear in pictures that crop out the body with its vices and passions and distill the subjects into a length of craniums. Peale was not fully satisfied, however, with his truncated images of the nation's leaders. Intrigued by taxidermic possibilities, and feeling that subsequent generations would benefit from seeing these wise men in perpetuity, he contemplated the idea of exhibiting the embalmed corpses of a select group of Founding Fathers. Nevertheless, as their friend, he was concerned about the distorting effects of preservation upon facial features. He finally abandoned the idea in favor of painting portraits of the revolutionary heroes and other leading citizens, an immense project that he undertook with characteristic zeal.[10]

The tool kit that Peale placed at his feet in his self-portrait is a reference not only to his taxidermy skills but also to the mechanical tinkering that occupied him throughout his life. Peale experimented with copying machines, telescopic rifle sights, and smokeless stoves, and he built a prosthetic hand for a member of the Pennsylvania state legislature. Devoting a portion of his galleries to machines, models, and mechanical drawings, Peale made use of these presentations to instruct visitors about the mechanical arts. He was a tireless proponent of new technologies that aimed to improve the comforts of the citizenry and he instilled in his students the idea that each of them had the aptitude to invent a useful device. Calling on Americans to experiment, he declared, "[man] can produce by the labor of his hands wonderful works of art, and with the knowledge of the lever, the screw, and the wedge, he can make machines to lessen his labour, and multiply the conveniences of life."[11]

The museum's mechanical marvels remain unseen in Peale's self-portrait, and another wonder is also partially hidden from view. The artist lavished attention on the flocks and luminaries portrayed on the left side of the picture, but he covered up the right side with the red curtain he raises with his right hand. Beneath the sumptuous fabric is a tantalizing glimpse of the tree-trunk limbs of the mastodon. Having been denied the federal subsidies that he had pleaded with Congress to give to his museum, Peale is resigned to be a man of business. The price of general admission to the galleries was fifty cents, with an additional charge of fifty cents to see the mammoth. Even the deep sense of pride that must have accompanied the artist's creation of the commemorative self-portrait did not persuade him to unveil there the jewel of his collections. Instead, the crafty impresario fashioned an image that is an enticing advertisement for the main attraction, and the animal that American naturalists had named the Great Incognitum would remain just that until the extra fee was paid.

Those who bought a ticket to advance past the drapery saw looming above them a noble and majestic being that embodied America's natural immensity. The mastodon's inconceivable dimensions flummoxed even the high-handed Georges-Louis Leclerc, the count of Buffon. The French author of a massive natural history encyclopedia and the foremost European authority on the subject, Buffon was also one of North America's most persistent detractors. He infuriated the United States' scientific community by maintaining that North American animals and plants were pitiful remnants of stock that remained vigorous in Europe. Thomas Jefferson led the charge against Buffon's calumnies, sending to Buffon lists of measurements of America's largest mammals, as well as the disintegrating pelt of a moose reported to be seven feet tall. The bones from the mire at Masten's farm became Jefferson's greatest weapon in his battle with his French adversary. In *Notes on the State of Virginia* (1787), the nation's defender remarks archly that the "skeleton of the mammoth (for so the incognitum has been called) bespeaks an animal of five or six times the cubic volume of an elephant, as Mons. De Buffon has admitted."[12]

As he worked out his composition for *The Artist in His Museum*, the aging Peale knew that his own extinction was also imminent. He organized his canvas as a painterly tour de force that would not only wrap up his career with a flourish but also communicate in the same language as the wonders of his museum. A daunting pictorial assignment for a self-taught painter, the representation of the deep and narrow Long Room required Peale to rely on a mechanical device to accurately plot the perspective. He also asked his son, Rubens, to assist him by making a preliminary study of the gallery. In homage to techniques perfected by Filippo Brunelleschi and other Renaissance masters, Peale pulls back the curtain to show the science of art. The painting is an exercise in seeing that asks the viewer to discern its proficiency with the mathematics of measurement and the triangulations required in surveying and engineering. Evidence of his characteristic unity of purpose, the artist's mapping of the room aligns with the geographical reconnaissance of collections gathered from across the United States and from distant territories in the West.[13]

The Artist in His Museum aims, however, to transcend both the craft of painting and the kinds of scientific observations that organized the museum. It is the extravagant sweep of the painting's crimson drape that signals that a program of Art is beginning. The curtain conveys the farewell of the revolutionary old guard—Peale, Jefferson, Adams would be dead before the end of the decade—and sets the stage for the nation's second act. It also alerts citizens that an act of imagination would be required of them. Indeed, the United States in the 1820s was more of an idea than a reality, a matter of fantasy and projection. At a time when many Americans feared that the country would not survive the loss of its founders, *The Artist in His Museum* insisted that it was not the revolutionary generation, but rather invention itself that lay at the heart of the national project. Peale opens his hand toward the viewer and, straddling a space between artistic and scientific vision, the old saddlemaker passes on the stewardship of his great American hall of wonders to the experimenters of the future.[14]

Peale died in 1827, five years after he painted his honorary self-portrait. His sons carried on the work of the museum for a time, but financial difficulties ultimately forced them to close its doors. Peale's collections were dispersed during the 1840s and 1850s, and much of the material ended up in P. T. Barnum's American Museum in New York. When a fire gutted Barnum's establishment during the Civil War, great banks of the specimens from Philadelphia were destroyed. Despite the demise of Peale's museum and the loss of many of its treasures, in the decades after the artist's death the spark of interest that he had fanned for so long combusted into an all-consuming American passion for science and mechanical invention. Citizens were still uncertain about the place of the fine arts in republican society, but they embraced science and technology as improving enterprises that would serve every member of the body politic and the nation's democratic institutions. Suddenly, a diverse and geographically fragmented population that argued about almost everything was bound together by the belief that the people of the United States shared a special genius for scientific discovery and invention. Significant numbers of Americans began to make their living in science or technology-related professions as engineers, cartographers, pharmacists, astronomers, and as teachers of geology, botany, and chemistry. Everyday people—lawyers, hat makers, and mill owners—took up scientific pastimes, scouring the countryside in search of botanical and geological specimens and piecing together pipes and springs in home workshops. They also devoured scientific journals and crowded into meteorology lectures at lyceum halls in Boston, Buffalo, and Pittsburgh. In their parlor cabinets, they lovingly assembled collections of shells, butterflies, and crystals.[15]

A portrait from 1839 of an anonymous "man of science," salted with intriguing clues about America's ardor for science and its applications, introduces one of the new generation of experimenters (fig. 3). Dressed in a black frock coat, starched shirt, and cravat, the unidentified subject of the painting sits in a workshop or laboratory where he is surrounded by an assortment of scientific objects. A globe on the left side of the canvas, its shape echoing the curve of the sitter's broad forehead, speaks of his absorption with geography. On the floor beside his chair, a pair of unfurling maps and a sprawled drafting compass suggest that the man of science has interrupted his geographical work just long enough to sit for his portrait. Is he a surveyor for one of the many nineteenth-century federal mapping projects? Or a speculator in local real estate with a special interest in the landscape visible through the room's deep-set window? Certainly, breakneck changes in the shape and size of the United States encouraged antebellum citizens to seek out newly published maps and to base their strategies for new enterprises on the information they contained. In the period when *Man of Science* was painted, the United States admitted Arkansas and Michigan to the Union; considered annexation of the new republic of Texas; shuffled the contours of its Iowa, Wisconsin, and Oklahoma territories; and argued with the British over its northern borders. Emblems of the country's expansion, the painting's globe and maps

link the man of science to the community of scientific researchers who were busy making inventories of the flora, fauna, and mineral resources of recently acquired land.

The window on the left side of the canvas, flanked by a thermometer that records fluctuations of temperature, establishes a visual connection between the work of the laboratory—assembling data, analyzing, and making predictions—and the explorations taking place outside it. Lying next to the man of science, a spotted hound, poised for a specimen-gathering expedition, turns his muzzle attentively toward his master. Hunting dogs were indispensable partners in the creation of zoological and botanical inventories; they helped naturalists locate not-yet-identified specimens and protected their owners from predators during forays into wilderness areas. Introducing the theme of retrieval, the sitter's four-footed protégé is also a reminder of contemporary scientific scouting efforts well known to American viewers and readers. In 1838, the year before the portrait was made, the United States Exploring Expedition set sail with a squadron of six ships to search the world for zoological, botanical, and mineralogical discoveries. In the same year, John Torrey and Asa Grey released the first volume of *Flora of North America*, a compendium of plants that drew on specimens provided by a network of amateur "botanizers" from the far reaches of the country. In 1838, John James Audubon published the final volume of his ornithological survey, the monumental *Birds of America*. In each of these enterprises, visual representation played a critical role in the recovery of scientific quarry, creating from ephemeral specimens a lasting image inspired by the idea of fidelity to nature.

While geography and assessments of the natural world are the subjects of the left side of the canvas, the right side is dominated by human contrivances aimed at harnessing nature's power and making it useful. The two apparatuses on the table are a universal furnace and a Woulfe bottle. The universal furnace, which produced the intense heat required in the production and separation of certain chemicals, and the many-necked Woulfe bottle were used by chemists creating new compounds. Both were pieces of equipment that Harvard lecturer John Webster advised for properly outfitted academic and amateur laboratories in his text of 1826, *A Manual of Chemistry*. Webster's equipment list recommended other items that also may have been on view in the man of science's laboratory, including "4 galvanic batteries, 1 Doz. Deflagrating spoons, Platina Wire, a large Brass Lamp Furnace . . . a Pepys Mercural Gasometer, a Best Double barrell'd Table air pump, a Galvanometer, Condenser & Parallel plates, a chemical Lamp, a metal Syphon," and a "Wollaston Reflective Goniometer."[16] If the room contained such contraptions, the portraitist—a person of limited painterly skills—was wise to banish them from his canvas. Instead, a few neat pieces of equipment offered his patron a flattering view of an orderly work space and allowed the artist to sidestep the technical vexations posed by a tangled array of tubes, bellows, glass bells, tongs, gears, and crystals.

Why did the man of science, a person of substance and learning, make his bid for immortality in a picture that forthrightly proclaims its maker's lack of expertise? Many contemporary observers scratched their heads over Americans' insatiable demand for pictures of themselves and their apparent lack of concern for the aesthetic quality of the works. In 1829 the artist John Neal described the public's taste for "wretched" pictures. "You can hardly open the door of a best room anywhere, without surprising or being surprised by, the picture of somebody, plastered to the wall and staring at you with both eyes and a bunch of flowers," he complained. "People being much more ready to pay for a picture of themselves or their children than for a picture of anything else, the portrait branch is always overcrowded with adventurers and cheap workmen; outcasts and bankrupts from every other department of the business."[17]

Americans of the period were familiar with works of fine art through reproductions of paintings by European masters that were widely available in bookstores and libraries. Despite the cultivated example of these models, middle-class American buyers maintained a robust appetite for homespun pictures made by "adventurers and cheap workmen." Where fine paintings spoke a highfalutin language, unskilled works addressed their subjects in an easy vernacular that soothed citizens who were self-conscious about their nation's remoteness from the elegant art salons of Europe. Amid the democratic jockeying of the Jacksonian period, amateur portraits presented an egalitarian face to the world that disregarded "aristocratical" art conventions, while they simultaneously insinuated a place for their patrons in the ranks of the respectable well-to-do. Expediency was the order of the day in a land of rapidly shifting circumstances, and the vagaries of fine art took a back seat to the practical concerns of science.

Americans were devoted to scientific advances, but many in society were excluded from such researches. Although the nation's women were energetic contributors to scientific endeavors, many of their accomplishments went unrecognized. "Women of science" were almost entirely absent from paintings and popular imagery. Scientific authorities, by definition male, held that females were unstable creatures dominated by emotions and instincts—incapable, therefore, of the objectivity and decisiveness demanded by scientific work. In *The Artist in His Museum*, Peale's depiction of a Quaker woman expressed the opinion of her gender that prevailed throughout the nineteenth century (fig. 4). Standing in the middle of the Long Room, the good lady reacts dramatically as she comes upon the great mastodon. She raises her hands in astonishment and seems to back away in fright to the edge of the avian diorama behind her. The woman is all flightiness and vulnerability; if we were closer, we might hear her excited twittering. Peale

FIG. 4
Charles Willson Peale
The Artist in His Museum
(detail, fig. 2)

was a liberal thinker who advocated for the education of women, but his condescending portrayal of the female museumgoer was just a polite version of rougher treatment elsewhere. Aware of the prejudices that surrounded them, most science-minded women thought it expedient not to speak of their interest. A well-known editor explained the facts of the matter: "If an unfortunate female should happen to possess a lurking fondness for any special scientific pursuit she is careful (if of any social position) to hide it as she would some deformity."[18]

The female citizens of the United States insisted that they, too, had assumed responsibility for the nation's Experiment, and many determined that they might participate in science by turning household work into a scientific enterprise. Sisters Catharine Beecher and Harriet Beecher Stowe, the authors of a bestselling homemaker's guide, *The American Woman's Home; or, Principles of Domestic Science* (1869), identified the home as the woman's laboratory

FIG. 5

Unidentified artist
The Kitchen. From Catharine Beecher
and Harriet Beecher Stowe, *The American
Woman's Home; or, Principles of Domestic
Science* (Boston: H. A. Brown, 1869)

and exhorted their readers to undertake their kitchen tasks and other domestic duties with systematic precision (fig. 5). The writers cast their net broadly in their book, covering a multitude of topics, such as "Scientific Domestic Ventilation," "The Ganglionic System," "Evils of Hot Bread," and "Phlegm." Not convinced, the nation's influential scientists, males all, dismissed "domestic science" as twaddle. Writing to a friend and fellow scientist in 1829, the celebrated botanist Amos Eaton complained about Emma Willard. An eminent educator and the founder of a prestigious seminary for young women, Willard had approached Eaton to discuss her wish to write a "kitchen chemistry" and ask him for advice. "Mrs. Willard," Eaton groused, "has been harassing [*sic*] me these two years about a Kitchen Chemistry. She thinks she could write a good one, and is asking me for suitable books to aid her. My repeated reply is in my usual abrupt manner with her. I tell her she is totally incompetent, and that I can direct her to do nothing."[19]

With the leading positions and research subjects in North American botany, geology, and zoology already spoken for by male scientists, many women decided to leave the planet to men and instead apply their talents to the celestial spheres.[20] The accomplishments of many female scientists centered in the field of astronomy, and even their male adversaries maintained that study of the stars and planets was naturally suited to the heavenly virtues of the "gentler sex." Because astronomy's territory was unlimited, it was impossible to exclude women from the field entirely. In the immense astronomical surveys underway in the United States during the nineteenth century, female "assistants" made vital contributions that required decades of painstaking observations with telescopes to calculate and document the position, size, and spectra of stars. Although female astronomers were paid extremely low salaries, their willingness to work for subsistence wages was an important factor in the high productivity of the projects. The director of the Harvard College Observatory, Edward Pickering, explained, "Many of the [female] assistants are skillful only in their own particular work, but are nevertheless capable of doing as much good routine work as [male] astronomers who would receive much larger salaries." Happily, those circumstances meant that "three or four times as many assistants [could] thus be employed."[21]

One woman in the United States gained wide recognition for her scientific work. When the astronomer Maria Mitchell discovered a comet in 1847, she became famous across the United States and in

Europe; the king of Denmark awarded her a gold medal for astronomy (fig. 6). The American public's opinion of Mitchell was divided, however. Women were cautiously thrilled by her achievements, while men wondered who would wash the dishes if the nation's homemakers were to devote themselves to such pursuits. Mitchell went on to direct the astronomy program at the newly founded women's college, Vassar, where she continued mapping the night skies for decades. Among the first astronomers to use photography to document sunspot activity, Mitchell was also an outspoken representative for women's ability to achieve in science. The inscription on the medal she received from the king of Denmark conveyed a poignant message to Mitchell and her sisters in astronomical endeavors: "*Non Frustra Signorum Obitus Speculamur et Ortus* (Not in vain do we watch the setting and rising of the stars)."[22]

Many of those who found themselves barred from the halls of science discovered that the door to its practical applications had been left ajar. Busy mothers, freedmen, and recently arrived immigrants learned that they could apply to the Patent Office with a new device and join the hallowed ranks of American inventors. Because patent application forms usually did not require applicants to list their gender, race, or country of origin, the patent system was open to an egalitarian community of tinkerers. In 1790 the United States had passed its first patent law, which stated that a patent could be awarded to "Any useful art, manufacture, engine, machine or device, or any improvement therein not before known or used." Allowing inventors to own their creations for a period of time, America's patent laws, in the words of patent holder Abraham Lincoln, "added the fuel of interest to the fire of genius."[23]

FIG. 6

Unidentified photographer
Maria Mitchell and Students Using Small Telescopes at the Vassar College Library
about 1878. Special Collections, Vassar College Libraries

Mark Twain extolled to the transformational effect of the patent system years later in *A Connecticut Yankee in King Arthur's Court* (1889). The narrator, Hank Morgan, remarks, "The very first thing I did in my administration—and it was on the very first day of it, too—was to start a patent office; for I knew that a country without a patent office

FIG. 7

Unidentified artist
Interior of the U.S. Patent Office,
Washington, D.C. From *United
States Magazine* 3, no. 4,
October 1856

and good patent laws was just a crab, and couldn't travel any way but sideways or backwards."[24]

The Patent Office building in Washington, D.C., was constructed on the plot that the city's original designer, Pierre L'Enfant, had set aside for a great national cathedral or a pantheon of American heroes. Instead, the massive building constructed on the site was a secular place of honor, a "Temple of Invention" that showcased the creative talents of average citizens. The Patent Office required inventors to prove through drawings and small models that their inventions were new and useful, and by the 1850s more than one hundred thousand people came each year to see the designs that filled every shelf of every cabinet in every gallery in the structure (fig. 7). Observers lingered reverently before the models of Eli Whitney's cotton gin, Cyrus McCormick's reaper, and Elias Howe's sewing machine, devices that had transformed American society and earned a lasting place for their makers in the annals of American heroes. The overwhelming majority of the models and drawings on view, however, represented only tiny improvements on existing technologies and practices. Nevertheless, visitors to the Patent Office left with the memory of the thousands and thousands of objects that declared that anyone anywhere could be an inventor. By midcentury, people began to carry out their daily routines with a razorlike focus on what they did and how they did it. Milking stools, thimbles, and streetlights became objects of fascination for citizens imagining patentable devices that would make a task easier or more efficient. In an environment where some patent holders were earning a fortune with rudimentary mechanisms, even the common privy merited hours of contemplation.[25]

Every patent model on display at the Temple of Invention had taken a unique route to Washington, D.C. Some mechanisms represented a maker's flight of fancy; others were propelled by years of grinding toil. Hard work was the driving force that brought Henry Blair's planters to the hallowed halls of the Patent Office (fig. 8). A free black from Montgomery County, Maryland, Blair in 1834 applied for and was awarded a patent for a seed planter; two years later he took out a second patent for a cotton planter. Blair had boldly entered

FIG. 8

Unidentified artist
Patent drawing for Henry Blair's
seed planter, 1834.
U.S. Patent and Trademark Office

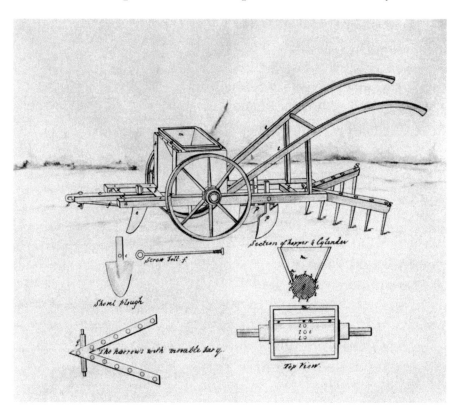

one of the few arenas of American enterprise where the referees of success and failure were not cued by race. Although many free blacks held skilled trade positions, state laws and local ordinances prevented them from ascending any further on the ladder of success. Many freedmen worked as laborers on farms and it was to a field task that Blair applied his inventive mind. Among the most burdensome and time-consuming of farm jobs was the proper planting of seeds, a critical element in determining the ultimate health of a crop. Perhaps it was as he mounded corn and cottonseeds on harrowed rows that Blair's imagination came to life. He pictured an apparatus that deposited seeds at the right spacing and depth and moved quickly through the fields; he may also have imagined plantings finished well before dark and the leisure hours to be enjoyed with the time his planter saved.

The models of Blair's planters at the Patent Office represented the victory of ingenuity over the social injustices of its time. Samuel F. B. Morse's telegraph model, on the other hand, represented a bitter defeat. Morse had worked for years to become a successful artist, having declared as a young man his aspiration to "rival the genius of a Rafael, a Michael Angelo, or a Titian."[26] In a self-portrait of 1812–13, Morse presented himself as an emissary of a dawning golden age of the arts in the United States (fig. 9). He wears the stately robes of London's Royal Academy, where he was a student at the time, and his fluid handling of skin tone, hair, and the texture of fabrics testify to his emerging skill. Morse belonged to a cadre of painters, novelists, and poets who shared Charles Willson Peale's belief that accomplishments in the arts were part of the United States' inventive work and the vector for visualizing its fragmented parts as a whole. The American people, however, took a stubborn pride in their no-nonsense outlook and were deeply suspicious of the illusionistic trickery of fine paintings. They also recoiled from art's historical alliances with Europe's decadent princes and popes. Colonel John Trumbull, one of the nation's most esteemed artists and a revolutionary hero, summed up the prospects of an artist in the United States when he was approached by a young painter who sought his advice. "You had better learn to make shoes," the colonel replied flatly. "Better to make shoes than to become a painter in this country."[27]

Even a half-century later, in 1859, an article in *Harper's New Monthly Magazine* confirmed that the United States remained a dismal environment for artists:

The elegant arts have existed among us rather as potted exotics imported from abroad, baubles to amuse the idle, luxuries to delight the rich, and, as such, awakening no real sympathy in the hearts of the people. The artist walks among us as a man apart, a solitary, a dreamer; misunderstood, unrecognized in the great working hive of society. Bookman looks askance at the ingenious handicraft; Hardfist despises the flaccid muscle and velvet palm; timorous Respectability has a horror of superfluous hair; venerable Conscientiousness is not sure but that the making of graven images and likenesses of things on the earth is contrary to Scripture.[28]

Determined to defy the odds against his success, Morse completed several years of study at the Royal Academy, painted portraits of the most important citizens of the day, and exhibited a collection of remarkable landscapes and history subjects. Still, he was unable to secure reliable patronage, and his financial circumstances were dire. In the 1840s, he surrendered. Calling art a "cruel jilt," Morse declared that he regretted the many years he had spent learning to be an artist. "I wish that every picture I ever painted could be destroyed," he announced. "I have no wish to be remembered as a painter."[29] Morse learned soon enough that the people of the United States would be happy to oblige him.

Between his painting assignments, Morse had tinkered in his studio with a prototype of an electromagnetic telegraph (fig. 10). He applied the visualization skills he had honed as an artist to see in his mind's eye how a telegraphic device would look and operate. Rummaging among the simple objects he found around him—a canvas stretcher, a rule, scraps of wire—he created a reliably functioning model of a telegraph and a code of "dot and dash" symbols. Inventors had experimented with the telegraph for years, but Morse was the first to come up with a working apparatus that was both easy to use and economical. Impressed by his demonstrations of the device, Congress in 1844 awarded Morse a grant of thirty thousand dollars to build a telegraph line from Washington, D.C., to Baltimore. The first message of his long-distance telegraph system was a question: "What hath God wrought?"

By the end of the 1840s, telegraph lines connected New York, New Orleans, and Chicago, and almost twelve thousand miles of wire

FIG. 10

Unidentified artist
Morse's Original Telegraph Apparatus.
From Samuel I. Prime, *The Life of
Samuel F. B. Morse, LL.D., Inventor of
the Electro-Magnetic Recording Telegraph*
(New York: D. Appleton, 1875)

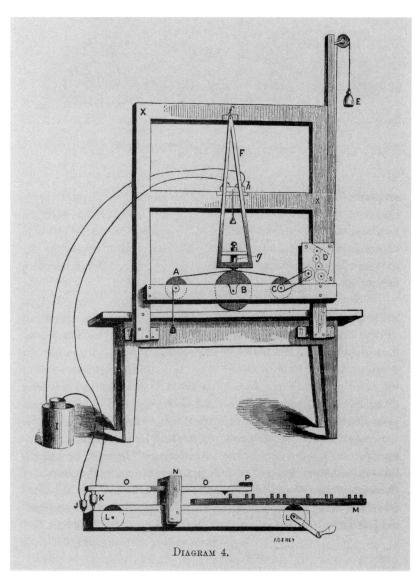

DIAGRAM 4.

had been strung; a few years later, that number had doubled. Morse's telegraph relayed information with unprecedented speed, volume, and regularity, and through its currents workers in the United States' communications, transportation, and commerce networks shook hands. Distant sweethearts exchanged endearments, steamboat captains kept informed about sandbars upriver, and farmers swapped news about the sizes of their potato crops. Public acclaim for the telegraph's unifying effects on the nation was so widespread and sustained that Americans struggled to find original ways to express their admiration. Boston physician William Channing hailed it as the "nervous system of this nation . . . its wires spread like nerves over the surface of the land, interlinking distant parts, and making possible a perpetually higher cooperation among men." He concluded, "By means of its life-like functions the social body becomes a living whole."[30] The popular *Yankee Doodle* magazine paid homage with an illustration that played with the subject of Morse's former occupation. Titled "Professor Morse's Great Historical Picture," the image replaced the artist's grand canvases with a telegraph wire strung across a wide-open landscape (fig. 11). The caption beneath the image declared, "Yankee Doodle expressed himself much pleased with the unity of design displayed in this great national historical work of art."[31]

Even years after the telegraph became a routine feature of their lives, Americans exalted it as one of the greatest wonders in an age of wonders, and they recognized Morse as a national hero. By the time of his death in 1872, their recollections of his outstanding contributions as an artist had been swept aside by the runaway success of his invention. One of those who never forgot Morse's artistic gifts was Rembrandt Peale, a son of Charles Willson Peale and his successor at the Philadelphia museum. Perplexed by the United States'

FIG. 11

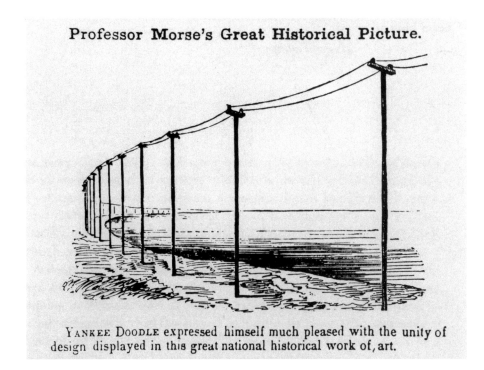

Professor Morse's Great Historical Picture.

YANKEE DOODLE expressed himself much pleased with the unity of design displayed in this great national historical work of, art.

obsession with technological invention and its neglect of art, Peale pondered the circumstances that had estranged the enterprise of the artist from those of the scientist and the mechanical inventor. All shared a quest, he believed, to understand the order of nature and reap the benefits that promoted individual and national well-being. Peale reminded his countrymen, as his father had before him, that art and science were often one and the same. With considerable pride and a little hope, he pointed out that American artists could take credit for many of the nation's most celebrated inventions. "It was a portrait painter, *Robert Fulton*, that gave us the power of steam navigation. It was a portrait painter, *S. F. B. Morse*, that devised the magic electric telegraph. It was a portrait painter, *C. W. Peale*, that first made porcelain teeth for himself and a few friends. And, I, though a portrait painter, lighted the first city with gas. This is no boast, but may be accepted as atonement for the practice of a luxurious art, which is now beginning to be appreciated."[32]

Democratic Time

HE DECLARATION OF INDEPENDENCE CLAIMED FOR the people of the United States the right to pursue happiness and the freedom to define how and when they would pursue it. The American Revolution, when "thirteen clocks were made to strike together," was a revolution in the conception of time.[33] As citizens threw off their obligations to foreign masters and set about using their hours as they saw fit, Democratic Time dawned over the land. The new Time represented the value of the minutes, days, and weeks of individuals empowered by democracy, expanding social mobility, and economic opportunity. During the first half of the nineteenth century, as the nation shifted from the seasonal rhythms of an agrarian economy to the accelerated tempo of

factory manufactures and integrated national markets, Democratic Time also came to stand for the promise that the nation's bounty was there for the taking and that the rightful takers would be they that got there first.

As Americans became people in a hurry, foreign visitors wrote about the unsettling influence of Americans' quest to do more, faster. Michael Chevalier, a French economist who arrived in the 1830s to investigate the nation's remarkable advances in communication and transport, reported that the United States was a country where "one lives a hundred fold more than elsewhere; here, all is circulation, motion and boiling agitation. . . . Experiment follows experiment; enterprise follows enterprise."[34] Such criticisms of the nation's hurried ways provoked indignant responses, including that of J. N. Bellows, writing in 1843 for *Hunt's Merchants' Magazine*. Bellows challenged "our calumniators" to remember that "every freeman in this country is part of the government; that he has to decide great questions daily." He pointed to the Herculean labors that awaited. "We have the land to clear, canals to dig, rail-tracks to lay, waterworks to finish . . . [and] cannot do everything today. Give us time."[35]

Although inventor Eli Terry could not conjure up time itself for the buyers of his products, he was determined to furnish them with a cheap and readily available way to keep track of it (fig. 12). Terry was one of a group of Connecticut clockmakers who pioneered the fabrication of inexpensive clocks using mass-production techniques in water-powered mills. During the first two decades of the century, the clock industry became one of the first and most important businesses to adopt the standardized parts system that started in American armories and spread quickly to other enterprises. The people of the United States were fiercely proud of what was known around the world as the "American System." With the proliferation of interchangeable-parts manufacturing, the land of *E pluribus unum* became a nation of a new creed—"out of one, many." Mass-production methods created a cornucopia of new products to buy, from wheat harvesters and sewing machines to steam printing presses and baby carriages.

Clockmakers at the beginning of the nineteenth century assembled individual clocks when customers ordered them, but Terry speculated that the low cost of his factory-made pieces would entice crowds of customers. In 1807 he accepted an order for four thousand clocks, an

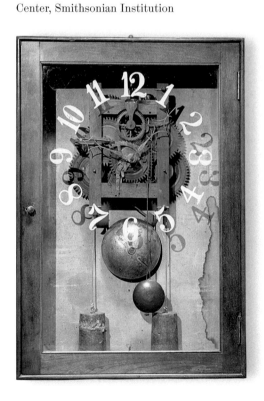

FIG. 12

Eli Terry
A mass-produced box clock, about 1816.
Division of Work and Industry, National Museum of American History, Behring Center, Smithsonian Institution

amount widely thought to be preposterous; when he finished the order in three years, he became a national hero. Rushing to expand upon Terry's success, other clockmakers contributed useful innovations that not only further reduced the already low price of factory clocks but also made them more dependable. Helping distribute the timepieces around the country, itinerant peddlers made sales pitches to rural schoolteachers and dairy owners, and to lumbermen who lived in the recesses of the forest.[36] An English visitor to the United States remarked in 1844 that even the remotest pioneer cabin in the hinterlands had a clock. "Wherever we have been in Kentucky, in Indiana, in Illinois, in Missouri, and . . . in every dell of Arkansas, and in cabins where there was not a chair to sit on, there was to be sure a Connecticut clock."[37]

An 1836 portrait of a schoolteacher, also the minister for a small town, pictures the well-tuned mechanisms of an egalitarian nation that produced abundant comforts for its citizens (fig. 13). Dressed in a sober dark suit with a high, stiff collar, Bartholomew Van Dame of Lee, New Hampshire, sits in an elegant Federal chair. As he consults a pair of books—one on the painted table in front of him and one in his lap—the sitter presents his profile to the artist. With his inkwell and pen at hand, Van Dame works in a space illuminated by a single lighted candle. Reigning over the Scripture-versed educator's studies is a "banjo clock," which hangs on the wall in the middle of the picture. The banjo clock was the brainchild of an American inventor, Simon Willard, who took out a patent for his device in 1802. Serving as a compact and economical substitute for the cumbersome tall case, or grandfather, clock, the banjo clock was a favorite of American customers. Most were made of fine hardwoods and brass embellished with painted and gilt decorations, a parlor adornment both practical and beautiful. Fancier models incorporated panels with painted landscapes or patriotic scenes, thermometers, and alarm bells.

In a democratic and rapidly shifting society, distinction in the quality of a clock or library was a way for citizens to assert their place in the social body. Americans were keen to show that their political freedoms did not preclude cultivation or access to the refinements of life, and they expounded regularly on the subject in sermons, election speeches, and poems. Horace Greeley, editor of the *New York Tribune*, wrote that the nation's new manufacturing system put into "the common possession of the people the enjoyments, the luxury, and the elegance, which in former

times were the exclusive privilege of kings and nobles."[38] However, while handsomely crafted clocks, silks, and fine furniture were widely available in the early decades of the nineteenth century, a fine painting by an accomplished artist remained hard to come by.

Bartholomew Van Dame's portrait provides a workmanlike outline of his face, figure, and the features of the room where he sits, but its handmade quality is out of kilter with the sitter's apparent prosperity and refinement. With its strange marriage of naive technique and genteel subject matter, Van Dame's portrait evokes the cultural grinding of gears that accompanied Americans' embrace of Democratic Time. Under the country's emerging scheduling regimes, which placed a premium on efficiency and productivity, the benefits of a career as a merchant or butcher accrued to the credit side of the ledger. A career as an artist, on the other hand, promised neither regular renumeration nor respectable social standing, making it difficult to reconcile a profession in painting or sculpting with the time value of money.

An undertaking that many Americans associated with the tyrannies and decadence of the Old World, the mastery of art called for years of specialized training in draftsmanship, modeling, composition, and color theory, with only a remote possibility that it would lead to financial success. Many hopeful artists, unable to afford the cost of advanced instruction in Europe, became self-taught entrepreneurs who hedged their art-making time with other professional activities. Those at the higher end of the economic scale had positions as postal clerks, bankers, and physicians and dabbled in painting or sculpting on the side. Artists with competent drawing skills could earn a decent living as engravers of banknotes, maps, and popular prints. The less gifted became traveling salesmen who added portrait painting to an assortment of skills and wares, shifting from decorating signs and carriages or selling clocks to painting portraits—until better prospects came along.

Davis posed Van Dame in a profile view, a format that allowed the artist to avoid the technical vexations of a full-face portrayal yet represented a step up from the common cutout silhouette. Showing off the decorative skills of a country craftsman, Davis lavished attention on the floorcloth's floral pattern and the trompe l'oeil wood grain of the writing table. The calligraphic flourishes below the image are made in the Italian Hand, an elaborate script taught by New England writing masters and also used for birth and baptismal documents and other family records. The embellishments in the picture enfold Van Dame in the long-established artisanal and family traditions of rural New England even as its communities were awakening, like Rip Van Winkle, to a new conception of time.[39]

While the maker of Van Dame's portrait was a country artist painting an elegant sitter, Francis Edmonds was an accomplished painter who made country bumpkins his specialty. Edmonds had studied the works of the Old Masters during a year in Europe and was a member of the prestigious National Academy of Design in New York. Aimed at the educated and prosperous city dwellers who were the buyers of his paintings, Edmonds's pictures featured rustic types puzzling over questions asked by census takers, tippling in the woods, and admiring the wares of traveling peddlers. His painting from 1857 of a country romance, *Time to Go*, captures the differing conceptions of time that segmented the citizenry at midcentury (fig. 14). Beneath a utilitarian pine clock that shows the time as almost midnight, a courting couple is lost in conversation while a fuming father rattles the stove grate. The cracks in the wall and the room's scuffed floorboards hint at his deferred maintenance projects and bills to be paid. The father's work

time and the couple's love time collide at an hour when "healthy, wealthy, and wise" citizens—the early-rising and early-to-bed folk that Benjamin Franklin applauded—had long been slumbering. A popular song of the period, "On Top of Old Smokey," warned that even the lovestruck needed to be mindful of the hour in the land of opportunity. The song's pokey suitor laments, "On top of Old Smokey / All covered with snow / I lost my true lover / For a-courtin' too slow."

For the upwardly mobile viewers who saw *Time to Go* at the National Academy as part of a polite circuit of city activities, the painting offered a backward glance at the places they had left behind. Small-town people dithering away the time, the trio in *Time to Go* was out of tempo with prevailing ideals of industriousness. In the decade and a half before Edmonds executed his painting, Americans had been busy indeed; they fought and won a war with Mexico, annexed California and a vast tract of southwestern territory, outfitted a fleet for Commodore Perry to sail to Japan, rushed for California gold, and invented the safety pin, the power rock drill, the elevator, condensed milk, and the sewing machine. In the United States, it was always time to go.

James Goodwyn Clonney was another artist who made paintings that were rich in allusions to old and new ways of managing time. A commentary on women's responsibilities in Democratic Time's regime, Clonney's midcentury *Mother's Watch* catches a woman dozing while a pair of mischievous offspring pilfers her watch (fig. 15). Mother seems to have nodded off while consulting Scripture, a testament to the level of engagement she brings to that worthy endeavor. Taking advantage of her moment of repose, her two sons clamber on a table to play with her watch. The glee on the boys' faces suggests that their mother does not permit them to handle her delicate timepiece when she is awake. With the guardian of the hearth asleep on her watch, the scruffy duo has the run of the place.

During the nineteenth century, females were expected to be the regulatory element of the American population, and napping was not on the agenda. Women were charged with the emotional, moral, and intellectual nurture and support of their children, husbands, and extended families; duties included serving regular hot meals and maintaining an attractive, wholesome home environment. In 1858, Harriet Beecher Stowe ridiculed the nation's impossible expectations of womanhood when she endowed the heroine of her *Atlantic Monthly* serial—published the next year as the novel *The Minister's Wooing*— with ideal attributes. "She shall scrub floors, wash, wring, bake and brew, and yet her hands shall be small and white; she shall have no perceptible income, yet always be handsomely dressed; she shall have not a servant in her house, with a dairy to manage, hired man to feed, a boarder or two to care for, unheard of pickling and preserving to do,—and yet you commonly see her every afternoon sitting at her

FIG. 15

James Goodwyn Clonney
Mother's Watch
about 1850, oil on canvas, 27 × 22 in.
Collection of Westmoreland Museum
of American Art, Gift of Mr. and Mrs.
Norman Hirschl

shady parlor-window behind the lilacs, cool and easy, hemming muslin cap-strings, or reading the last new book."[40]

With so many tasks to accomplish in a day, it is no wonder that Clonney's dear lady has fallen asleep. Still, the painting shows little sympathy, mocking the good intentions signaled by her index finger resting pointedly on her Bible. She may have been seeking heavenly comfort, but the imps in her parlor represent coming retribution. It was Stowe's sister, Catharine Beecher, who warned her countrywomen of the divine wrath that awaited those who neglected their duties: "A woman is under obligations to arrange the hours and pursuits of her family, as to promote systematic and habitual industry; and if…she refrains from promoting regular industry in others, she is accountable to God for all the waste of time consequent on her negligence."[41]

As increasing numbers of Americans moved to time-disciplined work habits related to industrial jobs, it was assumed that mothers' work in the domestic sphere would also conform to new models of time efficiency. By the 1850s, pocket watches were mass-produced, making it easier for the ladies of the house to always be aware of the lateness of the hour. Sadly, crying babies, thick roasts, and kitchen gardens tended to foil even the best-laid plans for exact scheduling. The mothers of the United States were expected to manage two opposing worlds, one made up of the spontaneous ebbs and flows of family routines and the other composed of segmented and quantifiable collections of minutes. A vision of a recruit overcome by battle fatigue, *Mother's Watch* is also a description of young, energetic male citizens ready to seize time.[42]

Lilly Martin Spencer, a prominent Ohio artist and mother of thirteen children, knew a thing or two about time management. The income from the sale of her paintings was the primary means of support for her family while her husband moved through a series of occupations and projects. Although Spencer's household was highly unconventional in a period when middle-class women did not have careers, her paintings focused on home scenes in which women did the cooking and sweeping up as men rested from the labors of enterprise. Spencer's painting, *Listening to Father's Watch*, 1857, shows a bearded man holding a gold watch up to the ear of a delighted child (fig. 16). Handsome in a rich dressing gown that indicates the end of the workday, the paterfamilias enjoys a relaxing interlude. He introduces his golden-haired child—it is not clear if it is a "he" or a "she"—to the joys of timekeeping. The child's face lights

FIG. 16

Lilly Martin Spencer
Listening to Father's Watch
1857, oil on academy board, 16 × 12 in.
Currier Museum of Art,
Gift of Henry Melville Fuller

up at the sound of the ticking clock as the father gazes tenderly, as if recognizing the preciousness of the fleeting days of childhood.

The mother is absent from this scene of domestic harmony, but her influence is implicit in the well-tended appearance of man and child and the cozy parlor where they sit. Indeed, her authority is paramount, for we know her to be the artist who made the painting. Spencer often used her husband and children as models for canvases like this one, and by working for a living she questioned the prevalent mores that compelled women to confine their activities to the domestic sphere. Her career as an artist also confronted the culture's pinched ideas of what constituted time well spent. Created by a career woman supporting her family, *Listening to Father's Watch* also speaks of the measured pace of generational time and the enduring satisfactions of family life.

As Spencer and other bold females were carving out space for themselves in what were traditionally male occupations, others were using their home activities as springboards for inventive activities. At midcentury increasing numbers of women were submitting patent applications for timesaving devices for homemakers. The celebrated successes of inventor heroes such as Eli Terry and Samuel F. B. Morse, as well as news of the considerable profits earned by those who sold their patents for minor innovations to entrepreneurs, galvanized women to incubate patentable ideas in their kitchens. Useful tools bubbled up in their imaginations—dough makers, cake stirrers, and stove blackers that shaved minutes or even hours off tedious jobs. At the Centennial Exposition in Philadelphia in 1876, a display of female inventions in the Women's Pavilion showed the fair's hundreds of thousands of visitors that patent-holding women were no longer an anomaly in the United States.[43]

Nineteenth-century women were outnumbered by men in the inventions of machines, but they were no strangers to the operation of machinery. When factory manufacturing emerged in the 1820s and 1830s, females were among the first to take up positions as factory "operatives." Men's labor was needed on the family farm, which was no longer the self-sufficient economic unit that it had been decades earlier. With the integration of national markets and the increasing availability of agricultural machinery and consumer goods, farmers required cash to make such purchases. The farmers' wives and daughters became important sources of needed revenue that allowed men to remain in agriculture. In his *Essays on Political Economy* (1822), Philadelphia entrepreneur Mathew Carey declared that by working in factories wives and children of farmers could "gather up fragments of time, which would otherwise have been inevitably lost." He also speculated that it was "probable that the profits of their labor were nearly equal, perhaps superior to the profits of farming."[44] With women's earnings, farmers could buy timesaving machinery, such as the popular reaper, patented by Cyrus McCormick. The inventor capitalized on Americans' recognition of the time value of money when he introduced the convenient "installment plan" that allowed buyers to make small down payments and complete the purchases through monthly installments.

Because the sequential, repetitive tasks of factory work required a synchronized workforce, factory owners adopted strict timekeeping systems to regulate their employees' activities. A schedule provided for operatives by the Lowell Mills of Massachusetts in 1851 laid out the parameters of the workday, which was organized through the ringing of bells (fig. 17). Bell towers were among the mills' most prominent features and rang at least a dozen times a day. Ringings woke workers, called them to breakfast and ended breakfast; the bells rang two minutes before the start of work, at the start of work, at the beginning and end of break time, at the end of the workday, and at the start and end of dinner. "Mill girls" came to loathe the sound of the bells and often complained about the incessant interruptions. A poem in the *Factory Girl's Garland*, a newspaper published by female operatives, described how it felt to be ruled by ringing. "The Factory Bell / Loud the morning bell is ringing / Up, up sleepers, haste away; / Yonder lists the red breast singing / but to list we must not stay. . . . Quickly now we take our ration / For the bell will babble soon; / Each must hurry to her station, / There to toil till weary noon."[45]

In a painting of 1871, *Old Mill (The Morning Bell)*, Winslow Homer, one of the United States' most successful artists, took up the subject of the women who worked in the mills. He chose to set his scene at the start of an industrial workday (fig. 18). Before midcentury, Americans viewed mills as places where respectable women could make important contributions to the nation's industrial transformation and also earn a good income. By the time Homer painted his canvas, native-born farmwives and their daughters, proudly featured in popular images and literary publications twenty years earlier, had long been absent from the mills. Immigrant and desperately poor women replaced them at the looms, the only takers for work that offered the barest sustenance. Their wages were often given as credits in the company store, which indebted them to their employers and prevented them from seeking better pay elsewhere. Walking the plank toward a structure that is dark and dilapidated, Homer's attractively dressed factory "belle" is commanded by a tolling that represents the despotism of the new order.[46]

TIME TABLE OF THE LOWELL MILLS,

To take effect on and after Oct. 21st, 1851.

The Standard time being that of the meridian of Lowell, as shown by the regulator clock of JOSEPH RAYNES, 43 Central Street.

	From 1st to 10th inclusive.				From 11th to 20th inclusive.				From 21st to last day of month.			
	1stBell	2dBell	3dBell	Eve.Bell	1stBell	2d Bell	3d Bell	Eve.Bell	1stBell	2dBell	3dBell	Eve.Bell
January,	5.00	6.00	6.50	*7.30	5.00	6 00	6.50	*7.30	5.00	6.00	6.50	*7.30
February,	4.30	5.30	6.40	*7.30	4.30	5.30	6.25	*7.30	4.30	5.30	6.15	*7.30
March,	5.40	6.00		*7.30	5.20	5.40		*7.30	5.05	5.25		6.35
April,	4.45	5.05		6.45	4.30	4.50		6.55	4.30	4.50		7.00
May,	4 30	4.50		7·00	4.30	4.50		7.00	4.30	4.50		7 00
June,	"	"		"	"	"		"	"	"		"
July,	"	"		"	"	"		"	"	"		"
August,	"	"		"	"	"		"	"	"		"
September,	4.40	5.00		6.45	4.50	5.10		6.30	5.00	5.20		*7.30
October,	5.10	5.30		*7.30	5.20	5.40		*7.30	5.35	5.55		*7.30
November,	4.30	5.30	6.10	*7.30	4.30	5.30	6.20	*7.30	5.00	6.00	6.35	*7.30
December,	5.00	6.00	6.45	*7.30	5.00	6.00	6.50	*7.30	5.00	6·00	6.50	*7.30

* Excepting on Saturdays from Sept. 21st to March 20th inclusive, when it is rung at 20 minutes after sunset.

YARD GATES,

Will be opened at ringing of last morning bell, of meal bells, and of evening bells; and kept open Ten minutes.

MILL GATES.

Commence hoisting Mill Gates, Two minutes before commencing work.

WORK COMMENCES,

At Ten minutes after last morning bell, and at Ten minutes after bell which "rings in" from Meals.

BREAKFAST BELLS.

During March "Ring out".......at....7.30 a. m..........."Ring in" at 8:05 a. m.
April 1st to Sept. 20th inclusive.....at....7 00 " " .... " " at 7.35 " "
Sept. 21st to Oct. 31st inclusive.....at....7.30 " " .... " " at 8.05 " "
Remainder of year work commences after Breakfast.

DINNER BELLS.

"Ring out"...... ....12.30 p. m........."Ring in".... 1.05 p. m.

In all cases, the *first* stroke of the bell is considered as marking the time.

B. H. Penhallow, Printer, 28 Merrimack Street.

FIG. 17

B. H. Penhallow
Timetable of the Lowell Mills,
Lowell, Massachusetts, 1851.
Baker Old Class Collection, Baker
Library Historical Collections,
Harvard Business School

During the last decades of the century, the quest for efficiency permeated every aspect of American life. Industrial managers sought ways to apply the techniques of the American system of manufactures—the process of mass-producing standardized parts that would be assembled into finished products—to human labor. A new professional, the "efficiency expert," entered the American workforce with the charge of maximizing the productivity of the employees of client corporations. The most sought-after nineteenth-century efficiency expert was Frederick Winslow Taylor, who claimed that "scientific management" could make any industrial task more efficient and, therefore, more profitable. Taylor analyzed in great detail the body movements of laborers chosen as average representatives, breaking their routines into a series of actions that he timed with a stopwatch.[47]

Based on his determinations of the average worker's productivity under optimized conditions on an average day, Taylor made recommendations for changes that would speed performance, and he created a precise choreography for the action of arms, legs, feet, and hands when shoveling, lifting, pushing a wheelbarrow, and other tasks. Taylor believed that the success of industrial corporations depended on their ability to separate the "brainwork" from the heavy-lifting jobs; he also espoused the idea that decisions about laborers' movements should be in the hands of management. Recommending piece-rate wages for workers, Taylor maintained that his system empowered laborers by giving them a stake in their own productivity. Taylor, however, used the most athletic and experienced workers in his efficiency studies, leading employers to set a pace that was impossible for average workers to maintain. This practical and ethical failure represented the heart of his system.

Taylor's observations of workers' motions were the foundation of all his teachings, and his visual studies had contemporary parallels. In 1883, the year he introduced the stopwatch to his examination of workers' movements, the photographer Eadweard Muybridge was invited to the University of Pennsylvania to undertake a project devoted to the

visual documentation and analysis of humans and animals in motion
(fig. 19). Muybridge had started his photographic study of motion in
1872, when the railroad tycoon Leland Stanford commissioned him
to determine whether all four of a racehorse's feet were ever off the
ground at the same time. Using a bank of twelve cameras and an elec-
trically controlled mechanism for releasing the shutters, Muybridge
documented the motions of his patron's winning Abe Edgington and
other trotting champions. The locomotion captured in the images and
their trainlike sequence dovetailed appropriately with the source of
the immense wealth that allowed Stanford, a former grocer, to participate in the "sport of kings."[48]

Muybridge's photographic studies at the University of Pennsylvania
extended his investigations into a wide range of human movements,
including the actions of men hammering, carrying heavy loads, and working with mechanical devices. Although the University of Pennsylvania's
commission of Muybridge's locomotion project did not define any specific
application, the university trustees believed that it would serve numerous
valuable scientific and artistic purposes. The project was consistent with
Philadelphia's long-standing role as a center of American science, technological invention, and fine arts; in addition, students at the university's
newly founded Wharton School of Business may have recognized the
pictures of men at work as a valuable study tool as they became aware

FIG. 20

Ferdinand Danton Jr.
Time Is Money
1894, oil on canvas, 16 $^{15}/_{16}$ × 21 $^{1}/_{8}$ in.
Wadsworth Atheneum Museum of Art,
The Ella Gallup Sumner and Mary Catlin
Sumner Collection Fund

of Taylor's concept that movement efficiency would bring "prosperity for the employee, coupled with prosperity for the employer."[49]

A painting of 1894 by Ferdinand Danton depicts the motivating power that was the driver of Democratic Time. *Time Is Money* shows an old wood door on which an alarm clock and a stack of paper currency hang from colored ribbons (fig. 20). Between the clock and the bills the word "is" is gouged into the wood. The legend "time is money" also appears in print on a mottled card below. It was the venerable Benjamin Franklin, who introduced the aphorism to Americans in an essay of 1748, "Advice to a Young Tradesman." Danton, already an accomplished painter when he executed *Time Is Money* at age seventeen, was himself the embodiment of the timely tradesman of Franklin's call to action.

Painted during a depression that followed the panic of 1893, during which hundreds of banks closed and one-fourth of the nation's railroads failed, *Time Is Money* is a masterful illusion that plays with the possibility of financial recovery. The paper bills it represents are not the common "greenbacks" of fluctuating and often questionable value, but rather the silver certificates that were redeemable in actual silver coinage. The adjacent clock, with its alarm hand set to go off at seven o'clock, reinforces the idea of value and industriousness—and, presumably, early rising. Nail holes in the wood boards behind it bring to mind wholesome manual labor, and the wood's well-used appearance speaks of a thrifty patching of materials. Nevertheless, the exact nature of the worn surface behind the clock and currency is uncertain. Is it a wall inset with a locked cabinet for hoarding treasures, or is it a door that opens to another place? With its painstakingly accurate construction vouching for the hours devoted to its creation, *Time Is Money* ultimately leaves a key component of the visualizing work to the viewer's imagination.[50]

Franklin's adage represents one of the earliest credos of the United States. His countrymen took the saying to heart, and by the early years of the century Americans were describing themselves as a "wide awake" people. In 1834 an English author wrote about the wakeful habits of the American body politic. Traveling from Baltimore to Norfolk, the writer noted, "Three fourths of the passengers had risen at four o'clock" even though "it was not until eight o'clock that we came in sight of Norfolk." The perplexed visitor mentioned the matter soon afterward to an American friend. "Ah, sir" said he, "if you knew my countrymen better, you wouldn't be surprised at their getting up at four o'clock to arrive at nine. An American is always on the lookout lest any of his neighbors should get the start of him. If one hundred Americans were going to be shot, they would contend for first place, so strong is their habit of competition."[51]

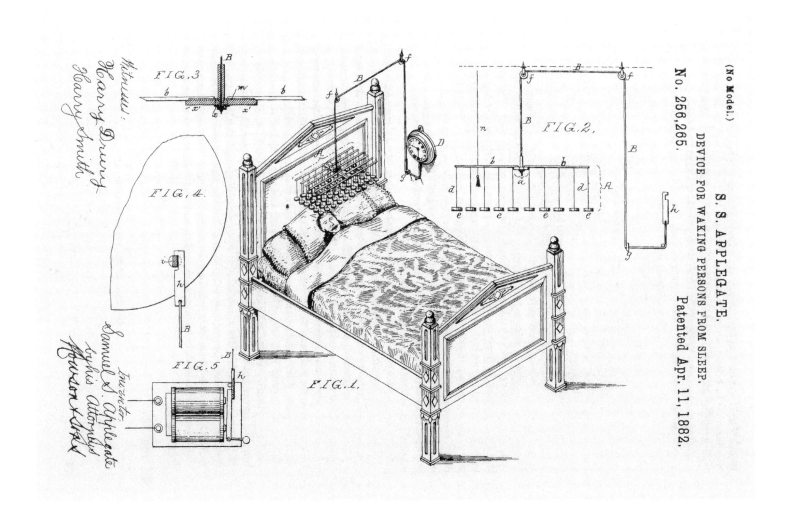

FIG. 21

Unidentified artist
Patent drawing for Samuel
Applegate's Device for Waking
Persons from Sleep, 1882.
National Archives and Records
Administration

Spurred by knowledge that many Americans were able to sleep through loud alarm-clock bells, Samuel Applegate, an enterprising inventor, in 1882 took out a patent for a device that would rouse even the most recalcitrant sleepers to rise and meet the day (fig. 21). "I suspend a light frame in such a position that it will hang directly over the head of the sleeper . . . whereby the frame is at the proper time permitted to fall into the sleeper's face. . . . The only necessity to be observed [in] the frame being that when it falls it will strike a light blow, sufficient to awaken the sleeper, but not heavy enough to cause pain."[52] It is not clear if Applegate's "Device for Waking Persons from Sleep" was targeted toward the heavy-lifting or the "brainworker" element of the United States' population, but the invention was premised on the idea that any American who remained asleep as a new day dawned would appreciate a timely knock upside the head.

Whatever I've been told of thy wonders is true!
All nature seems at once to rush on my view,
And, lost in the trance you occasion, I cry,
How stupendous the scene! What an atom am I!

"Lines Written at Niagara," *The Port Folio* (1807)

Niagara Falls

HREE YEARS BEFORE THE CENTENNIAL OF THE American Revolution in 1876, the popular magazine *Harper's Weekly* published an illustration that assembled a disparate crowd of Americans at Niagara Falls (fig. 22). *Niagara Seen with Different Eyes* shows the myriad visions that defined the celebrated cataract over the previous decades: the "eye of the tourist," "eye of sentiment," "eye of love," "eye of patriotism," and "eyes of practical possession, eyes of wonder, eyes of pleasure, / Eyes of fancy deep, eyes of toil and eyes of leisure." The image poked fun at the magazine's audience, for presumably every *Harper's* reader could identify with one of the eyes in the lineup. Indeed, during the nineteenth century Niagara Falls was a turbine for

FIG. 22

Arthur Lumley
Niagara Seen with Different Eyes.
From *Harper's Weekly*,
August 9, 1873

all sorts of American imaginings and the subject of a steady flow of artistic portrayals, scientific insights, and engineering innovations. With its immeasurable power derived from the various bodies of water that converged at the falls, Niagara emerged as a symbol of a nation founded on the idea of *E pluribus unum*. A visit to the falls bound millions of nineteenth-century citizens in a thrilling and patriotic experience. Even the Yankee in the *Harper's* picture—an American type famous for cryptic, monosyllabic utterances—waxes enthusiastic about the dazzling spectacle of Niagara. "Wa'al, I te yeu now, we kalkerlate that's some in the way o'water-power."[53]

Niagara Falls was the largest known waterfall in the world during the first half of the nineteenth century. The brightest jewel in the United States' crown of natural beauties, Niagara was marvelous, in part, because cultivated Europe had no counterpart. Every giftbook on

American scenery included a picture of it, and every person who could afford the cost of the journey made a trip there. Over the course of the century the manner of depicting and viewing the falls changed radically, prodded by shifts in ideas about nature and physical changes brought on by the development of the area around the falls. The aspirations of a divided citizenry mingled there as artists, scientists, engineers, factory owners, and carriage drivers offered their explanations of what Niagara was and to whom its beauty and energy belonged. For many Americans, Niagara represented an eddy in the rushing currents of Democratic Time, a place where they could pause to ponder the purpose of the nation and contemplate the timeless majesty of nature.

Science Unveiling the Beauties of Nature to the Genius of America, a drawing of 1814 by the British topographical landscapist John Barralet, made an auspicious prediction about the United States from the vantage point of a Niagara landscape (fig. 23). The left side of the image shows an allegorical America wearing a liberty cap, seated where the white rush of Niagara can be seen in the distance. She is surrounded by the botanical and zoological plentitude of the New World; sunflowers,

White-headed Eagle.

corn, squash, and melons spill over the ground at her feet, flanked by a curious American menagerie of rattlesnake, beaver, moose, and snapping turtle. As Niagara exhales its mists in the background, winged "Science" removes Nature's drape, revealing her four-breasted beauty. The felicitous pairs serve as Barralet's cheerful reminder of the extraordinary opportunities that awaited America's ingenious people and the twofold harvests that would follow well-directed science and art.[54]

Recently emigrated from Scotland, the artist, naturalist, and poet Alexander Wilson roamed the wild country near Niagara Falls in 1804 and drew from his memories there as he worked on his grand survey of American birds, the four-volume *American Ornithology; or The Natural History of the United States* (1808–14). For his picture of the bald eagle, the monarch of America's winged creatures, he created a vision of Niagara Falls as the setting (fig. 24). In a poetic description of the eagle, Wilson wondered at the grace of the bird that glided calmly above Niagara's magnificent "horrors."

High o'er the watery uproar, silent seen,
Sailing sedate, in majesty serene,
Now 'midst the pillared spray sublimely lost,
And now, emerging, down the rapids tost,
Swept the grey eagles; gazing calm and slow,
On all the horrors of the gulf below.[55]

Many of the specimens that Wilson gathered for his book ultimately landed in Charles Willson Peale's Philadelphia museum; Wilson also went there to sketch the birds that Peale had found and stuffed. It probably was at Peale's museum that Wilson found, lording over a bevy of glass-eyed ducks, robins, and gulls, the stuffed bald eagle that he used as the inspiration for his illustration in *American Ornithology*. Peale advised the museum's visitors that birds served as models of republican virtues, and Wilson's pairing of the bald eagle with Niagara Falls invoked the idea of the United States' freedoms and its growing power.[56]

Wilson made many drawings of the falls during his visit, one of which takes a perspective that looks up to the great vertical drop of Horseshoe Falls from below. Wilson's "View of the Great Pitch Taken from Below," engraved by John Barralet in 1810, beckoned to the discriminating readers of the *Port Folio*, a magazine devoted "to the Useful Science, the Liberal Arts, Legitimate Criticism, and Polite Literature" (fig. 25). The drawing brought viewers to the opening of the windswept, slippery cavern behind the sheet of Horseshoe Falls. Tourist accounts abounded with descriptions of the harrowing zone behind the sheet, where torrents of water dashed all who dared pass and dramatic barometric shifts snatched their breaths away. "View of the Great Pitch Taken from Below" appeared in the *Port Folio* before it was certain that all those who risked the trip would emerge from the cavern unscathed, challenging the polite perusals of its audience with an image of Niagara's crushing force.

The sights and sensations of Niagara tended to erode distinctions between scientific and artistic ways of thinking, and painted landscapes of Niagara in the early part of the century collaborated with topographies of the area. The citizens of the United States inserted maps into every aspect of daily life: decorative wall maps, magazine maps, surveying maps, gazetteer maps, and maps on playing cards and needlework samplers. Addressing young geographers in her

FIG. 25

John James Barralet after Alexander Wilson
View of the Great Pitch Taken from Below.
From *Port Folio* 3 no. 3 (March 1810). The
New York Public Library, Astor, Lenox and
Tilden Foundations

FIG. 26

George Catlin
Bird's Eye View of Niagara Falls
about 1827, gouache, 17⅝ × 15½ in.
Private collection

Geography for Beginners (1826), Emma Willard advised that geographical education was the route to getting ahead. A "good knowledge of [geography]," she explained, "will add to your respectability, by causing you to be regarded as a person of information."[57]

Accurate knowledge about Niagara Falls and its environs was one of the first lessons for a person of information in the United States, and intelligence about its features was conveyed in painted and printed landscapes as well as in maps. William Woodbridge, in his popular schoolbook *A System of Universal Geography* (1824), imparted the idea of a symbiosis between geography and art when he wrote that learning to read maps was like learning to draw a whole human body. "The novice in drawing first delineates individual objects or the several parts of the body," he said. "It is the business of a more advanced stage of his progress to draw even a single human figure; and it is not until he is master of the elements of the art, that he is permitted to combine a variety of objects into a group or a landscape and to imitate the coloring of nature."[58] Paintings and drawings of the Niagara landscape, then, served as geography tutorials that assembled maps' fragmented abstractions into a whole.

Like the United States itself, however, Niagara resisted assembly. Its multiple waterfalls, rising mists, intermittent rainbows, scattered boulders, and seething whirlpools conspired to fumble even the most accomplished artists; those who approached the falls without a plan were doomed. George Catlin tackled the problem in 1827 by painting Niagara as an aerial map and adding three-dimensional features to his cartographic scaffolding: houses, white-capped currents, travelers on roads, and a forest canopy (fig. 26). Other nineteenth-century pioneers in painting Niagara organized its geographical complexity through the conventions of the picturesque, a philosophy of landscape painting formulated by the British writers Sir Uvedale Price, William Gilpin, and others. Painters looking into the maw of the waterfall struggled to realize the picturesque formula, which called for compositional balance and variety and subject matter highlighting bucolic harmony, mild weather effects, quaint countryfolk, and, usually, ruminants. One of those who

FIG. 27

John Trumbull
Niagara Falls from Below the Great Cascade on the British Side
1808, oil on canvas, 24 7/16 × 36 3/8 in.
Wadsworth Atheneum Museum of Art,
Bequest of Daniel Wadsworth

wielded pencil, crayon, and brush to tame Niagara into a semblance of the picturesque was John Trumbull. A revolutionary hero and an esteemed painter of portraits and history scenes who journeyed to the falls in 1807, Trumbull knew how to manage a challenging campaign. As a colonel during the War of Independence, he drew military maps for George Washington and enlisted his topographical skills as he took stock of Niagara's complicated features.

Trumbull rallied picturesque ideals of pictorial harmony and variety with a mapmaker's quest for accuracy in *Niagara Falls from Below the Great Cascade on the British Side* of 1808 (fig. 27). Arching trees frame the scene on the left and right, and clearly delineated bands of foreground and midground progress in an orderly manner toward a hilly background. After the foreground's congenial introduction to Niagara—an esplanade where a British soldier strolls with a pair of fashionable ladies and an artist sketches in the crook of a tree—the serenity of the composition drops away suddenly. The falling water beyond appears as two white rectangles, one atop the other like a set of gnashing teeth, and a serpentine column of mist rising from the depths of the cascade hints at the destruction boiling below. Isaac Weld, a British author who visited the falls a decade before the artist gazed out from the British side, described the fierce energies that opposed Trumbull as he labored to compose a picturesque Niagara. "No words can convey an adequate idea of the awful grandeur of the scene at this place," he insisted. "Your eyes are appalled by the sight of the immense body of water that comes pouring down so closely to you from the top of the stupendous precipice." Weld wrote of "trembling with reverential fear" when he imagined, as every visitor did, falling "into the dreadful gulph [sic] beneath, from which the power of man could not extricate you."[59]

The artist Alvan Fisher traveled to Niagara Falls during the summer of 1820 and came upon a very different scene than the one Trumbull had encountered years earlier. When Trumbull journeyed to the falls, the roads were poorly maintained and fallen logs, deep mud, and washed-out sections made the region inaccessible to all but military men and the most determined pilgrims. As Fisher set up his paints and sketching materials on the embankment, he saw hundreds of fellow tourists and an assortment of newly constructed hotels in "the Front," a growing quarter-mile commercial zone that included taverns, souvenir booths, a curiosity museum, and hawkers selling guidebooks and rock specimens. The new ease of a trip to Niagara was one of the rewards of the frenzy of road building in the United States during the 1820s that established a network of gravel toll roads across important thoroughfares. The Erie Canal opened soon afterward, in 1825, allowing Niagara excursionists to float there in comfortable barges at a smooth four miles an hour. With *The Great Horseshoe Fall, Niagara* of 1820, Fisher linked his painterly observations of topographical features and landmarks with his impressions of the stunning ephemeral effects of the falls and the exhilaration they elicited in visitors (fig. 28).

Perhaps because the routes to Niagara were improving and tourism was booming, Fisher was careful to show in *The Great Horseshoe Fall, Niagara* that a visit to the falls was not for sissies. He also affirmed that it could not be confined within picturesque pictorial conventions. Introducing his panoramic view with parties of tourists

in the foreground, he shows a young woman who covers her ears to shut out Niagara's deafening roar; on the left a pair of men help a comrade over the bank as he reaches the top of a flight of treacherous wooden steps that lead down to the bank of the river. Contemporary newspaper accounts and travel narratives had described the infamous stairs in lurid detail, and the American public was steeped in tales of the physical and emotional ordeal they represented. Rickety and wet with spray, the stairway subjected those fit enough to attempt them to buffeting winds and crumbling rocks encrusted with sulfurous ooze. The worst part of the test lay at the bottom, on the path to the ledge behind the sheet, a perilous spot that every nineteenth-century American knew from engravings in books and magazines. It is possible that even Raphaelle Peale's trompe l'oeil *Venus Rising from the Sea—A Deception*, painted around 1822 during the early days of the national "sheet" mania, may have played off audiences' familiarity with the drenching thrills that awaited tourists behind the falls (fig. 29).[60]

The English traveler Frances Trollope described her excruciating efforts to advance behind the sheet: "We more than once approached the entrance to this appalling cavern, but I never fairly entered it, though two or three of my party did," she recounted. "I lost my breath entirely; and the pain at my chest was so severe that all my curiosity could not enable me to endure it."[61] Basil Hall, a retired Scottish naval captain who ventured behind the sheet seven years after Fisher painted *The Great Horseshoe Fall*, recounted: "Though I was only half an hour behind the Fall, I came out much exhausted, partly with the bodily exertion of maintaining a secure footing while exposed to such buffeting and drenching, and partly, I should suppose, from the interest of being in this scene, which certainly exceeds anything I ever witnessed before."[62] Fisher captures in his painting the relief and excitement of the stair-climber's return in the companions who rush to pull their friend from the brink. Silhouetted against the churning immensity of Niagara, the trio acts out a now-familiar story under a span of rainbow that appears above their heads.

Nineteenth-century Americans spoke of the trip down the Niagara staircase as a journey back in time that offered firsthand evidence of the geological forces at work in the region. The foremost science in the United States during the first half of the century, geology was one of the few scientific arenas in which the United States compared favorably to its European peers. Citizens took great pride in Niagara's rocky erosions, recessions, and sedimentary shifting, which allowed them to speak with authority about ancient time in a country where almost everything was brand new. A scientific handmaiden to religious beliefs, geology grew into a great national passion during the religious revival movement that swept the nation in the 1820s. While church pews across the country filled with salvation seekers looking for holy guidance, the newly inspired flocked to the shale and dolostone precipices of Niagara to search for signs of divine action on Earth. Guidebooks offered precise information about the mineralogical formations visible from the staircase at Niagara and touted the experience behind the sheet as an unforgettable lesson in God's power. Visitors read the

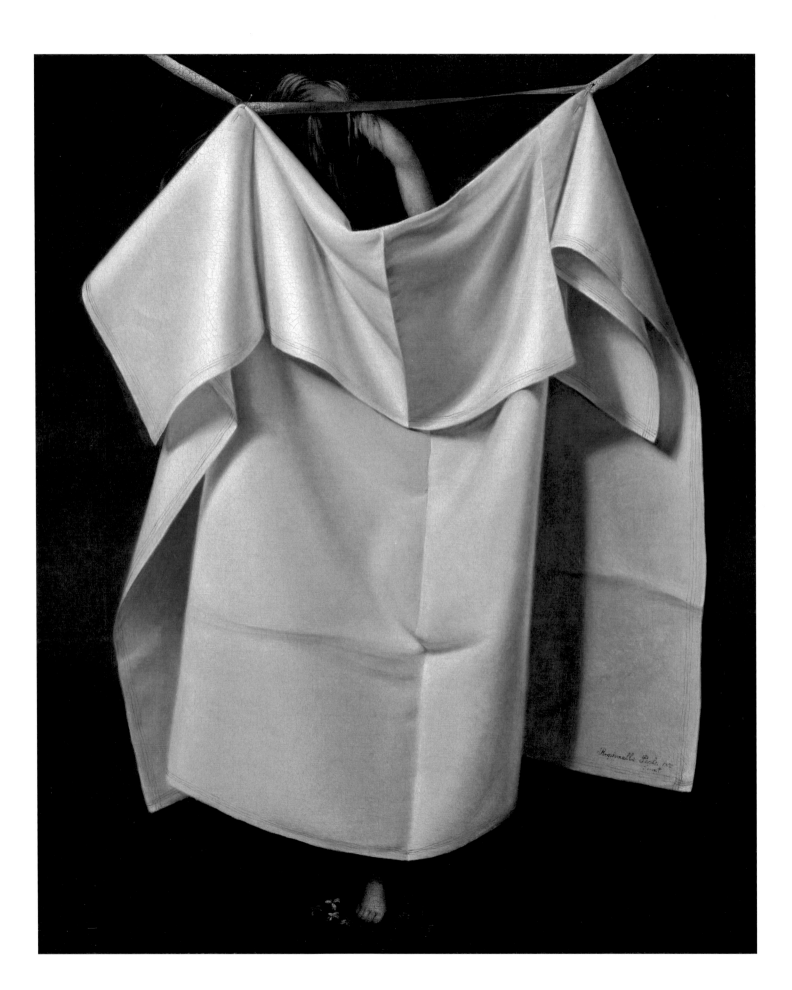

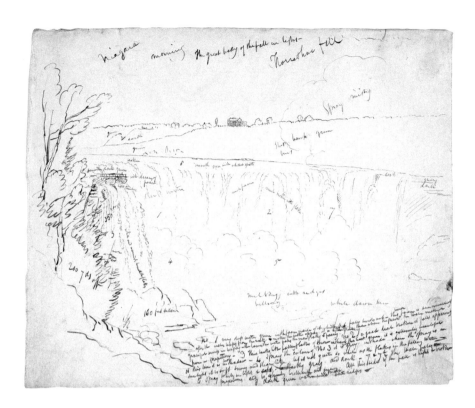

strata of the cliffs and great boulders along the ledge of the Horseshoe Falls as the tracings of the Omnipotent One.

Thomas Cole, a landscape artist who emigrated from England in 1818, was a devout student of geology and scripture. As part of his religious, geological, and artistic musings, Cole wrote about Niagara in his widely read "Essay on American Scenery" (1835). "And Niagara! That wonder of the world!" he exclaimed. "At our feet the floods of a thousand rivers are poured out—the contents of vast inland seas. In its volume we conceive immensity; in its course, everlasting duration; in its impetuosity, uncontrollable power."[63] The artist set out for Niagara in 1829 with plans to capture its "everlasting duration" on canvas. Shaped by his geological researches, Cole's preliminary studies of the falls emphasized disciplined observation and documentation. His drawings included not only the groupings of forms in the landscape but also the detailed notes about weather and geological elements of an experienced naturalist (fig. 30). In *Distant View of Niagara Falls*, a widely acclaimed painting that resulted from his sketches at the site, Cole laid out a landscape undergoing a turbulent transformation (fig. 31). Charcoal clouds gather momentum in an unsettled sky as the terrain ignites in the vermilions of autumn. The artist sets a pair of

FIG. 31

Thomas Cole
Distant View of Niagara Falls
1830, oil on panel, 18 7/8 × 23 7/8 in.
The Art Institute of Chicago,
Friends of American Art Collection

Indians in the line of sight to the falls, situating the country's original people as mediators between the viewer and this "wonder of the
world." After carefully delineating the features of the surrounding scenery, Cole shifted his painterly strategy as he addressed the falls itself.
He captured the water's steep drop and the mists, froth, and whirlpool
with rapid and impressionistic strokes, as if loath to commit Niagara's
wildness to the captivity of his depiction.

Cole derived many of his ideas about Niagara and its geology
from his friend Benjamin Silliman, professor of natural history and

chemistry at Yale (fig. 32). On rock-collecting expeditions, Cole and Silliman debated geological theories as they roamed the New England countryside picking up pieces of crystals and sandstone to elucidate points about volcanic activity and the action of water. Silliman was especially interested in uncovering signs of the great flood described in the Old Testament. He and the many eminent American scientists who had once been his students believed that the Deluge—God's great punishment for human wickedness—was an authentic historical event that could be traced by the countless signs that remained of its destruction. J. L. Comstock wrote about the Deluge in his *Outlines of Geology* (1834). "The effects of that grand and awful cataclysm are still to be traced in every country, and in nearly every section of . . . the globe," he advised. "Vast accumulation of rounded or waterworn pebbles, huge blocks of granite, and immense beds of sand and gravel are found in places where no causes now in operation ever could have placed them; and still that they have been moved is evident from the circumstances or places where they occur."[64] Comstock's description perfectly matched the landscape at Niagara, and the faithful contemplated the inundations of the falls as a vivid reminder of God's wrath.

Drawing from the copious geological observations recorded in his notebooks, his discussions with Silliman and other geologists, and descriptions in the Bible, Cole painted his vision of the aftermath of the Flood in an imposing, three-by-four-foot canvas (fig. 33). *The Subsiding of the Waters of the Deluge* of 1829 begins in the shadows of a rocky overhang from a vantage point that peers out to a sunlit morning. A denuded landscape is littered with signs of the cataclysm: a broken mast, a skull, and trees reduced to tangles of shredded bark. All signs of human habitation and the blanket of soil that nurtured its cultivations have been washed away, exposing the earth's rocky skeleton. Crags, outcroppings, and escarpments loom as warnings of the devastating finality of divine judgment. For Cole, an outspoken critic of Americans' materialistic orientation and their hasty destruction of the country's

FIG. 33

Thomas Cole
The Subsiding of the Waters of the Deluge
1829, oil on canvas, 35 ¾ × 47 ¾ in.
Smithsonian American Art Museum,
Gift of Mrs. Katie Dean in memory of
Minnibel S. and James Wallace Dean
and museum purchase through the
Smithsonian Institution Collections
Acquisition Program

scenic beauty, Niagara was a revelation of both the miracle of God's grace and the terrible consequences of incurring his displeasure.[65]

Frederic Church was Cole's sole student and the heir to whom he passed the mantle of an American landscape art that encompassed a moral and scientific vision of nature. Church was the United States' leading artist at midcentury, so it was not a question of whether he would devote his energies to a grand interpretation of Niagara Falls, but when.[66] Inspired by the current ideas of the German naturalist Alexander von Humboldt, who wrote about the "chain of connection" that linked natural systems across the earth, Church sought out the truths of landscape with the objectivity of a scientist and, at the same time, a deeply spiritual sensibility. His name was famous across America, and audiences jostling for space at the crowded openings of his exhibitions brought binoculars to be sure of a good look. When word spread in 1856 that Church was at Niagara making studies for an important painting, audiences nationwide waited with keen expectation.

Church laid the groundwork for his project as his master had taught him, making a variety of sketches from different vantage points around the falls in pencil, gouache, and oil (fig. 34). Traveling there on three separate trips in 1856, he compiled an exhaustive record of visual phenomena, including the wind's effect on the trajectory of spray, the gradients of blue and green in the current, and the texture of sedimentary formations. A contemporary art reviewer described Church's eye as a camera—a device revered in the United States as a truth-telling machine—but one enhanced by an artist's judgment of color and balance: "His eye, like every other man's, is a camera with a brain behind it," the critic explained. "But his brain gives him the power to transfer to canvas the vanishing forms and tints and shadows thrown upon the eye, unaffected by the medium through which they have passed, except in his selection, combination, and unification."[67]

FIG. 34

Frederic Edwin Church
At the Base of the American Falls, Niagara
1856, brush and oil on paper laminate, 11 ⅝ × 13 ¾ in., Cooper Hewitt, National Design Museum, Gift of Louis P. Church

Church embarked on his visual study of Niagara in the wake of a terrible disappointment that struck the heart of the nation in 1855. In that year the explorer David Livingstone, hunting for the source of the Nile in southern Africa, stumbled upon the great waterfall that he would name in honor of Queen Victoria. Livingstone described Victoria Falls—twice the height and breadth of Niagara—as a scene so resplendent that it "must have been gazed upon by angels in their flight."[68] Word of the discovery soon reached the capitals of Europe and the United States; measurements were taken, calculations made, and Victoria Falls was pronounced the largest waterfall in the world. The ascendancy of Victoria Falls was a humiliation for Americans, who felt that one

FIG. 35

John Rapkin
Niagara, United States
about 1851, engraving. From
Montgomery Martin, *Illustrated Atlas,
and Modern History of the World*
(New York: John Tallis & Co., 1851).
12 ⅝ × 9 ½ in. Castellani Art Museum
of Niagara University Collection,
Generous Donation from Dr. Charles
Rand Penney, partially funded by
the Castellani Purchase Fund, with
additional funding from Mr. and Mrs.
Thomas A. Lytle, 2006

of their most valuable possessions—and a strategic weapon in their
ongoing rivalry with Europe—had been stripped from them (fig. 35).
It is not known if Church's project was influenced by the discovery of
Victoria Falls, but the timing of his decision in 1856 to paint Niagara
was impeccable. National audiences were hungry for an affirmation of

the importance of the nation's iconic waterfall and embraced the subject of Niagara with renewed patriotic fervor.

In a painting completed in 1857, Church threw down the gauntlet (fig. 36). He swept away the botany, geology, and meteorology of his studies, the crowds and hotels at the falls, and the visual conventions passed down to him from Cole and previous masters. Paring out almost all but water from the scene, he left only a sliver of landscape in the distance as a life raft for the eye. The painting shows the view across Horseshoe Falls, the section that flowed over the ledge behind the sheet that had so terrified Wilson and Trollope years earlier. It is there that Church casts the viewer on the waters an instant before the plunge into the "awful gulph" below, creating a lover's leap that unmoors the sense of an individual self. Looking away from the painting, the viewer is surprised to see the walls, floor, and ceiling of the room where the canvas hangs—so convincing is the artist's cascading invention. Even the African usurper—larger, yes, but beyond reach in the forests of that distant continent—must yield to the force majeure of Church's genius vision of America's fountainhead.

A reviewer for the distinguished art magazine the *Crayon*, exclaimed: "We have never seen [water] so perfectly represented."[69] Art critics in London raved about the painting's "vast multiplicity of water effects—foam, flash, dark depth, turbidity, clearness, curling, lashing, shattering, and a hundred more."[70] The painting was indeed a discourse on the action of water, and its vision of overwhelmingly hydrological energy addressed a country newly transformed by water-powered innovations. Steam engines dominate modern histories of the nineteenth century, but for the generations living in the period when Church painted *Niagara*, waterpower was the principle agent of change. The water mill was the handmaiden of the famous innovations that Americans hailed as examples of their inventive genius, beginning with Oliver Evans's celebrated automated flour-milling process in 1785. Steam only began to provide the majority of manufacturing power after 1875; from Maine to Mississippi water-powered gristmills, sawmills, fulling mills, gunpowder mills, paper mills, and mills for making glassware, nails, and farm tools served as a unifying feature across a dispersed nation. Waterpower meant bread, construction materials, and everyday household conveniences; waterpower mills were often the first structures to be built in new communities, even before churches and schools.[71] Basil Hall, who

FIG. 36 ☞

Frederic Edwin Church
Niagara
1857, oil on canvas, 42½ × 90½ in.
Corcoran Gallery of Art,
Museum Purchase, Gallery Fund

FIG. 37

Thomas Doughty
*Mill Pond and Mills, Lowell,
Massachusetts*
about 1833, oil on canvas, 26 × 35 in.
Private collection

FIG. 38

Unidentified artist
Why Doesn't the Water Drop Off?
From Jacob Abbott, *Rollo's Philosophy:
Water* (Boston: Phillips, Sampson, 1855)

had quaked behind the sheet in 1827, related Niagara to America's river-harnessing projects when he likened its roar to a gristmill. "In hunting for similes to describe what we heard . . . we were quite agreed that the sound of the Falls most nearly resembles that of a grist mill, of large dimensions."[72]

Americans viewed falling water as the natural dynamo that created everything from cotton blouses to pocket revolvers. They enthused over its power not only in landscape art but also in pictures of water-driven factories, in books for children, and in lyceum lectures about hydrological action (figs. 37–39). Americans bragged about the nation's wealth of waterways, and as avid readers of atlases and almanacs that compared the United States' resources to those of Europe, they learned that in England and France the rich and powerful monopolized virtually all water-mill sites. What was jealously guarded in the Old World was available to all in the United States. American rivers, then, carried a message of political freedom and egalitarian, ever-flowing opportunities. Emblematic of God-given bounty, Niagara Falls was not harnessed until later in the century precisely because the United States possessed so many other sources scattered across the land.

While Church was working out the vectors of the raging currents at Horseshoe Falls, American readers were engrossed by images and

WHY DOESN'T THE WATER DROP OFF?—Page 17.

FIG. 39

Bass Otis
Interior of a Smithy
by 1815, oil on canvas, 50 ⅝ × 80 ½
in. Courtesy of the Pennsylvania
Academy of the Fine Arts,
Philadelphia, Gift of the artist

descriptions in a recently published book about hydrology (fig. 40). *Lowell Hydraulic Experiments* (1855) documented a series of more than 160 hydrological trials carried out during the 1830s and 1840s by James Francis, chief engineer at the famed Lowell Mills in Massachusetts. Because American science had not yet produced useful texts on hydrology, millwrights and mill owners of the period were forced to rely on European studies that were often unsuited to conditions in the United States. The owners of Lowell Mills charged Francis to gather new information. Using several kinds of waterwheels and turbines, he measured the flow of water of different depths over weirs and dams of varying heights. Because calculus and theory were not yet perfected, Francis relied on close observation and accurate visual documentation in his analysis of water flows. By eliminating the logjam of inadequate data and directing significant improvements in the hydraulic systems that

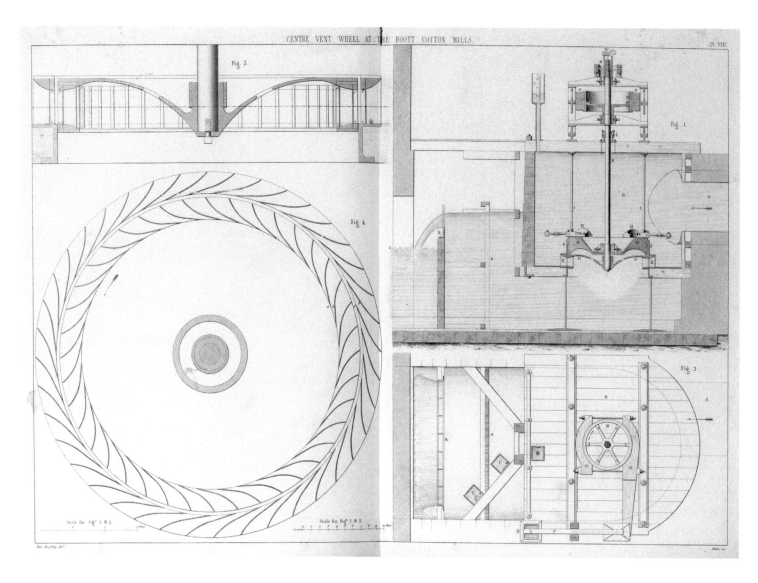

FIG. 40

Unidentified artist
Centre Vent Wheel at the Boott
Cotton Mills. From James B. Francis,
Lowell Hydraulic Experiments
(Boston: Little, Brown, 1855)

powered the factory, the Lowell experiments represented a critical
step in the nation's progress toward economic independence. Although
Lowell's river-harnessing theories had already been adopted in many
American factories, the publication of the elegantly bound *Lowell
Hydraulic Experiments* formally presented to the American public the
innovative vision that drove the Lowell Mills' twenty thousand horse-
power turbines and its commercial success.[73]

Always eager to keep pace with the enthusiasms of the American
public, the enterprising P. T. Barnum arranged to display his own
hydrological marvel at his American Museum in New York. Barnum
purchased a barrel-and-pumping contraption for two hundred dollars,
then posted an advertisement for the "Great Model of Niagara Falls

with Real Water." Barnum later wrote in his autobiography about ticket buyers' reactions to the "Great Model": "A single barrel of water answered the purpose of this model for an entire season; for the falls flowed into a reservoir behind the scenes, and the water was continually re-supplied to the cataract by means of a small pump," he remembered. "Many visitors who could not afford to travel to Niagara were doubtless induced to visit the 'model with real water,' and if they found it rather 'small potatoes,' they had the whole Museum to fall back upon for twenty-five cents, and no fault was found."[74]

If P. T. Barnum was a great showman, Frederic Church was an even better one. Viewers claimed that his "Great Picture" had instructed, exhilarated, and spiritually uplifted them; they left the exhibition hall convinced that the artist had distilled onto his canvas the very wellspring of America. The contemporary author Adam Badeau alluded to this upwelling of patriotic emotions when he called Church's *Niagara* a "true development of the American mind, the result of democracy, individuality . . . of the liberty allowed to all."[75] If *Niagara* was an embodiment of the nation, it is no coincidence that the scene is set at the edge of an abyss. The metaphor is unmistakable today, for hindsight shows us the struggles over slavery and states' rights that were pushing the country to the brink of the Civil War. Church finished the painting in the year when the Supreme Court galvanized both abolitionist and supporters of slavery with its ruling in the notorious Dred Scott case, deciding that it had no power to limit a state's authority over an individual's disposition of private property, including slaves. Part of a recent and accelerating series of events that spurred sectional hatreds—"Bleeding Kansas" and the caning of Charles Sumner in the United States Senate, among them—the Dred Scott decision hardened positions on both sides of the debate over slavery. Church captured the rising urgency of the time with his picture of a great precipice and, like a diabolical ferryman, used his paintbrush to conduct the eyes of the nation toward a place that allowed no turning back.

On the boundary between the United States and Canada, Niagara Falls had long functioned as a geographic and cultural gateway. At Niagara the right to own slaves ceded to Canada's laws against holding slaves. Niagara was also one of the few points on the American border where fugitive slaves could cross to freedom. Underground Railroad operators and bounty hunters converged there; the dangers of the

THE COMING MAN'S PRESIDENTIAL CAREER, à la BLONDIN.
Morro.——Don't Give up the Ship.

FIG. 41

Cartoonist possibly Jacob Dallas
The Coming Man's Presidential
Career, à la Blondin. From *Harper's*
Weekly, August 25, 1860

river represented the final hurdle in the harrowing ordeal of an escape as well as a last chance for bounty hunters to secure rewards of thousands of dollars for nabbing private property on the run. "The Fugitive Slave's Apostrophe to Niagara," a poem of 1841, calls out to the river as a divine rescuer:

> *Eternal Priestess, thine*
> *Is the pure baptism of the chainless free;*
> *Cool on this brow of mine*
> *Thy holy drops descend, as broad to me*
> *Unroll the temple-gates of meek-eyed Liberty!*

After 1848, when a suspension bridge was built over the river, the passengers of the Underground Railroad avoided the hazard of the currents by walking on the tracks above the river. Fear of oncoming trains and the dizzying height of the bridge caused many to falter midway and turn back. The intrepid Harriet Tubman, who escorted scores of fugitive slaves to safety in Canada, always carried a pistol with her to shore up the resolve of "passengers" who faltered just a few steps away from "Ca'naan."[76]

In 1860 an illustration in *Harper's* pictured the leader who soon would be hailed as the "Great Emancipator" negotiating the dangers of Niagara with a fugitive (fig. 41). The Republican Party's presidential candidate, Abraham Lincoln appeared in the magazine just a few months before the election of 1860. Outfitted in the tights and pantaloons of the professional "funambulist," Lincoln furrows his brow in concentration as he edges across a tightrope high above the river. He totes a wary slave on his shoulders and steadies himself and his cargo with a balance bar emblazoned, "Constitution." A sign in the background—"To the Whirlpool"—points out the consequence of a misstep. The *Harper's* image was referencing the fabulous exploits of Monsieur Blondin, the death-defying "Prince of Manila" who had braved Niagara on a tightrope the year before. Blondin's aerial feats—which included carrying his manager on his back across the span and cooking eggs on a stove above the water—were the talk of the nation, and the name "Blondin" quickly became synonymous with agility and aplomb in the face of grave danger. Famously gangly and awkward, Old Abe was the physical opposite of Blondin, but *Harper's* made a cheering prognostication when it compared him to the celebrated acrobat who never once stumbled: "The Coming Man's Presidential Career, *à la* Blondin," reads the hopeful legend at the bottom of the illustration.

The *Harper's* picture of the future president showed a man adept at surmounting obstacles; in fact, Lincoln had worked on a flatboat as a young man in Illinois and, along with repairing boats and hauling, learned how to extract his vessel from frequent strandings on sandbars and immersed debris. Most Midwestern captains accepted groundings as an unavoidable part of river travel in the shallow waters of the region and relied on shovels, plank wedges, and liberal applications of elbow grease to refloat vessels that were stuck. Convinced that a better way could be found, Lincoln came up with an idea for a flotation device and worked it into a model for a patent application that he submitted in 1849. His "Device for Buoying Vessels over Shoals" was aimed at the

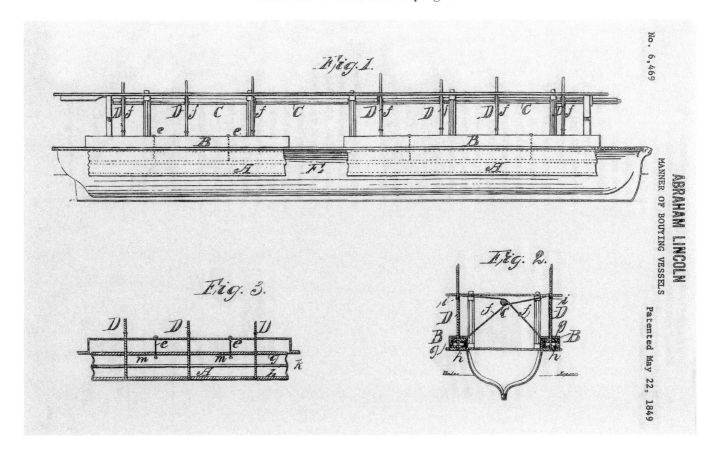

flourishing steamboat trade of the Midwest; it included a system of inflatable chambers that, when filled by an air pump, lifted vessels off sandbars and through shallow water (fig. 42). Although Lincoln's device was never manufactured, his interest in innovation served him well when he assumed command of the ship of state in 1860. After the secession of the Confederate states, he took full advantage of improved technologies—steamships, telegraph, railroad, and breech-loading rifles with telescopic sights—to preserve the union that he swore to protect as president.

Lincoln nevertheless faced looming shoals during the Civil War that had nothing to do with firepower or communication with his generals, but rather with white citizens' beliefs about the character and abilities of black Americans. Those convictions often were channeled in pictures of rivers; a typical example, "Way Down upon the Swanee Ribber," illustrated in *Harper's* in 1873, is evidence of the broad current of racism that flowed through the white person's imagination during the nineteenth century (fig. 43). Over the title of the minstrel song of 1851 by Stephen Foster, a black street fiddler—his sign reads, "Songs 5 cents"—dreams of happy days by the Swanee. The river provides a slow-moving backdrop for African Americans who are depicted whiling away the time with dancing and banjo playing. Implicitly, the picture asked its white audience: what is the place of such people in our go-ahead nation? Certainly, those who cared to look could find abundant proof that contradicted the image of lazy Negroes down by the "Ribber." Rivers were not only routes for countless ingenious slave escapes but also realms where both the enslaved and freedmen claimed a social space for themselves as shipwrights, boat captains, and inventors of maritime improvements.

FIG. 43

Richard Norris Brooke
Way Down upon the Swanee Ribber.
From *Harper's Weekly*, June 28, 1873

Benjamin Montgomery, a slave owned by Joseph Davis, brother of the Confederate president Jefferson Davis, made a habit of observing the steamboats that plied the waters near the Davis plantation (fig. 44). Skilled in architectural and mechanical design, Montgomery envisioned a change in the angle of the boats' propellers that would give them more thrust. He then devised an angled steamboat propeller that not only increased the average speed of the paddle wheelers but also reduced their fuel costs and provided a smoother ride for passengers. Both Joseph and Jefferson Davis recognized the importance of Montgomery's invention and, in one of the baffling contradictions of Southern slavery, diligently tried to have it patented in Montgomery's name. The attorney general of the United States ruled in 1858 that a machine invented by a slave, no matter how novel or useful, could not be patented. Still, during the Civil War the Confederacy adopted Montgomery's ingenious propeller for steamships in its navy.[77]

Hydraulic innovations transformed the United States during the first half of the nineteenth century, but the immense power of the falls at Niagara resisted significant changes until the 1880s. Artistic invention had been the project there, the challenge of portraying in original fashion a subject that had been worked over countless times before. Would-be developers proposed numerous schemes to harness the "foam and fury" of the falls, but public outcry prevented their implementation. One of Niagara's most famous defenders was Frederic Church, who helped organize a campaign to preserve the falls' beauty and banish the penny arcades and tawdry shops that marred the surrounding landscape. In a startling display of creative hubris, Church also advocated remodeling the falls themselves, maintaining that judicious changes would give them a more balanced appearance. "The natural formation of the rocks seemed to invite some artistic treatment especially by cutting channels for the purpose of forming picturesque cascades," he explained, envisioning improvements that "would not only greatly enrich and diversify certain portions, but also do much toward harmonizing the general effect."[78] Church's Olympian remodeling recommendations went unheeded, but he became a leading figure in the creation of a public park in the immediate environs of the falls in the 1880s, keeping at bay the wax museums and refuse dumps that threatened to engulf it.

Niagara Falls assumed center stage in an energy revolution that occurred in the United States in the 1890s, and it was at that pivotal

FIG. 45

George Inness
Niagara
1893, oil on canvas, 45 $\frac{1}{4}$ × 70 $\frac{1}{8}$ in.
Hirshhorn Museum and Sculpture
Garden, Smithsonian Institution,
Gift of the Joseph H. Hirshhorn
Foundation, 1966

moment that George Inness decided to create a series of paintings there. On commission to a railroad company, Inness had portrayed locomotive operations in the 1850s and was familiar with the topography of technological change as he looked out on the planned site of the largest hydroelectric power plant in the world. He sketched the falls as engineers, financiers, and inventors were debating the optimal locations and materials for an immense hydroelectrical project at Niagara. Above the protests of American preservationists who insisted that the cataract was a national treasure that belonged to the whole country, the Niagara Falls Power Company and numerous partners worked out the technological systems necessary for the generation, transmission, and use of electricity on a magnitude that had never been attempted. The project required not only the complete rethinking of earlier generations of turbines and motors but also the invention of a host of new generators, transformers, and transmitters that didn't exist yet. In 1890 workers began blasting through three hundred thousand tons of rock to excavate a discharge tunnel eighteen feet high and seven thousand feet long. One of the visionaries of the project was Nikola Tesla, who said that Niagara contained enough power to "light every lamp, drive every railroad, propel every ship, heat every store, and produce every article manufactured by machinery in the United States."[79] Newspapers in all parts of the country kept fascinated readers apprised of every aspect of the mammoth power station taking shape at Niagara Falls, the showcase of American ingenuity that was to open in 1895.

When Inness painted *Niagara* in 1893, the noise of drills, hammering on metal, and steam engines echoed across the gorge as the gargantuan tunnel next to the falls was nearing completion (fig. 45). Mechanics and engineers swarmed over the site as they assembled the largest turbines ever fabricated, including a huge, twenty-nine-ton double engine that soon would be lowered into the massive wheel pits in view, also the largest ever built. By 1893 public acclaim for the wonder and promise of Niagara's new technologies had drowned out the voices of the multitudes that deplored the project as desecration. H. G. Wells, when visiting the falls after the completion of the station, articulated the point of view that prevailed. "The dynamos and turbines of the Niagara Power Company ... impressed me far more profoundly than Behind the Sheet; [and] are indeed, to my mind, greater and more beautiful than that accidental eddying of air beside a downpour."[80] Tapping into the inventive energies swirling around Niagara in 1893, Inness released his *Niagara* from the conventions that had guided artistic portrayals for more than one hundred years. The painting offers no topographical tour, no information about a particular season or time of day, and the sailors, soldiers, brides, poets, and Indians have been dispatched. We discern the deluge of Niagara behind a sheet of aqueous colors and forms that set our imaginations adrift. A red flag in the foreground and a smoking industrial chimney float like signal buoys in the strange dreamscape of *Niagara*, Inness's nebulous vision of the United States at the threshold of illumination.

The Peacemaker

A PAIR OF NINETEENTH-CENTURY INVENTORS FROM western New York, C. M. French and W. E. Fancher, were hopeful as they mailed an application to the United States Patent Office for their "New and Improved Ordnance Plow" (fig. 46). As required by the patent examiners, they included a full description of the twofold purpose of their device. "As a piece of light ordnance its capacity may vary from a projectile of one to three pounds weight without rendering it cumbersome as a plow," they explained. "The combination enables those in agricultural pursuits to have at hand an efficient weapon of defense at very slight expense in addition to that of a common and indispensable implement. . . . Its utility is as an implement of the

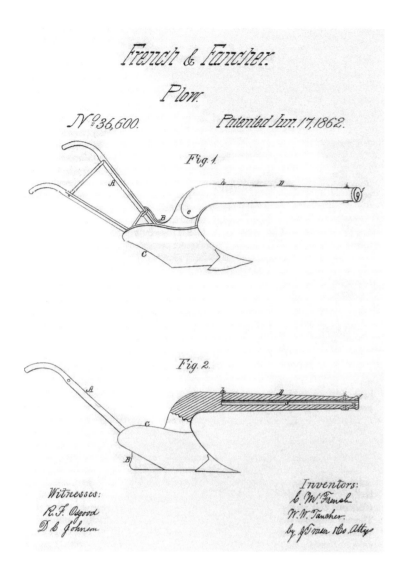

French & Fancher.
Plow.

N.º 36,600.   Patented Jun. 17, 1862.

Fig. 1.

Fig. 2.

Witnesses:
R. F. Osgood
D. C. Johnson

Inventors:
C. W. French
W. W. Fancher
by J. ..... Co. Att'y

twofold capacity described, especially when used in border localities subject to savage feuds and guerrilla warfare. As a means of repelling surprises and skirmishing attacks on those engaged in a peaceful avocation it is unrivalled." Little is known about the enterprising Misters French and Fancher, aside from the fact that they came from Waterloo, New York, a town in a region of small farms populated by farm families. Of these hardworking, self-reliant agrarians—the citizens that Thomas Jefferson hailed as the guardians of the nation's democracy—which ones did they envision as the customers for their ordnance plow? If the prospect of a market for a tilling and killing machine raised disturbing questions about the kind of nation that Americans had cultivated in the century since the Revolution, this was not part of the Patent Office's purview. The invention was deemed to be both novel and useful by the examiners and was awarded U.S. patent number 35,600.

Certainly, the United States was born with a gun in its hand, brought to life by what Ralph Waldo Emerson would describe years later as "the shot heard round the world." In April 1775 a British regiment marching to Concord, Massachusetts, to seize arms and ammunition collected by the rebellious Americans encountered a ragtag party of colonial Minutemen—farmers, dairy owners, and other common people who could muster with their muskets at a minute's notice. The War of Independence began when one of them squeezed the trigger of his weapon, a hayseed David confronting the Goliath of the British Empire. When war began in earnest, George Washington brought in the colonies' most proficient marksmen, calling for nine companies of expert riflemen to be raised from the backcountries of Pennsylvania, Maryland, and Virginia. The backwoodsmen arrived with their tomahawks and hunting knives and wearing buckskin breeches and moccasins decorated with porcupine quills. They also carried Kentucky rifles, homespun weapons that had been developed to suit the conditions of life in the borderlands. The rifles had long barrels that gave bullets deadly accuracy and grooved interiors that propelled the ball in a spiraling action, firing them at least twice the distance of a projectile shot from a musket.

On the other side of the battlefield, the reputation of the western riflemen preceded them, for tales of Americans' formidable gunning abilities had circulated in Europe for decades. An eighteenth-century Anglican cleric giving an account of their skills observed that "the great quantities of game, the many kinds, and the great privileges of killing make the Americans the best marksman [*sic*] in the world." A fellow Englishman agreed. "There is not a man born in America that does not understand the Use of Firearms and that well. . . . It is almost the first thing they Purchase and take to all the New Settlements and in the Cities you can scarcely find a lad of twelve years That [does not] go a Gunning."[81] Word spread among the British regiments that their opponents were crack shots who could drive a nail though a board—or shoot a man through the left eye—from a distance of two hundred yards. An English captain declared that "a Rifleman grows naturally behind every Tree and Bush on the Continent." In London, *Gentleman's Magazine* apprised readers that an easy victory against the "provincials" was not assured. "Their guns are rifled barrels, and they fight in ambush; five hundred provincials would stop the march of five thousand regulars," it warned. "And a whole army might be cut off, without knowing where the fire came from."[82]

Despite their top billing, the performance of the backwoods riflemen during the war was lackluster. They balked at taking orders from commanders, and their long-barreled guns were maddeningly unreliable; the infantry's smoothbore muskets proved to be more valuable than rifles because muskets could be loaded so much faster. Like their comrades from the backwoods, the regular troops showed great enthusiasm for entertaining themselves with gunning horseplay. They used their muskets to start fires, shoot at geese flying overhead during maneuvers, and kill and wound one another with stunning regularity in accidental shootings. An exasperated General Washington remarked on the problem in 1776. "Seldom a day passes but some [soldiers] are shot by their friends," he complained.[83]

It was not the Americans' expert marksmanship that ultimately prevailed in the War of Independence, but rather a fortuitous combination of skill in the art of ambush, hard-slogging work with muskets and bayonets and, most important, the coming together of the disparate communities of the colonies—people who usually viewed each other with suspicion and contempt—to cooperate for a common cause.[84] As the actual circumstances of war faded from American memory during the nineteenth century, however, the vital role of *E pluribus unum* in the victory also seemed to disappear. By the time Thomas Jefferson died in 1826, the gun had emerged in the public imagination as the agent of the United States' noble Revolution and the founding ideals of "life, liberty, and the pursuit of happiness." The artists, novelists, inventors and other creative citizens who were fashioning a national vision of what the United States was and should be increasingly identified the gun as the source of America's independence.

Nineteenth-century citizens believed that most Americans possessed a marksman's eye, a natural visual acuity that gave them special advantages in their business dealings and social exchanges. "[Americans] may be considered as peculiarly distinguished for possessing the faculty of sight in a great degree of accuracy and perfection," a Boston editor opined in 1815. "In the first place, there is that skill in gunnery which may be almost regarded as a national characteristic, from its simplest form as it appears in the quickness and accuracy of eye and certainty of fire of our millions of marksmen, up through every form of the art of artillery to the highest branches of practical engineering."[85] Given the abundance of self-congratulatory citizen accounts of American marksmanship and the wealth of gun-themed paintings that came by midcentury, it is curious that guns were mostly absent from early genre images that showed average people going about their everyday routines.

Many nineteenth-century American artists, including James Goodwyn Clonney, Francis Edmonds, John Lewis Krimmel, and William Sidney Mount earned respectable livings producing pictures of common folk; guns rarely figured in their works. Certainly, the high level of detail in paintings featuring guns endorsed the idea of a keen American eye, linking marksmanlike vision in art to the intelligence gathering that was considered to be a vital part of the nation's democracy. Americans maintained that clarity of sight and the ability to discern truth—particularly about one's fellow citizens—was of paramount importance for the members of a democratic electorate who needed to see and decide matters for themselves. Nevertheless, guns played only bit parts in works that described the amusing habits of cider makers, fiddlers, cornhuskers, and schoolteachers. When firearms did make an appearance, typically carried by hunters or "sportsmen," the people toting them were a far cry from the legendary riflemen of revolutionary days. The hunter was often cast as a bumbler or ne'er-do-well, someone who idled away his time chasing ducks and deer while decent people were hard at work.

In *The Sportsman's Last Visit*, painted in 1835, William Sidney Mount assigned to a hunter the role of the loser in a love contest (fig. 47). Dressed in a brown hunting jacket with his game bag slung over his shoulder, the sportsman scratches his head in consternation as he casts

FIG. 47

William Sidney Mount
The Sportsman's Last Visit
1835, oil on canvas, 21 ¼ × 17 ¼ in.
The Long Island Museum of American
Art, History & Carriages, Gift of
Mr. and Mrs. Ward Melville, 1958

a baleful eye at his rival. The other man, attired in a formal black suit and starched collar and clearly a person of refinement, addresses the sweet-faced miss who is the object of both men's affections. Lowering her eyes demurely, the young lady listens intently to the gentleman in black and pays no mind to her second caller. The targets of Mount's painting, of course, were the well-heeled and would-be well-heeled who attended exhibitions of his works at the National Academy of Design. They saw themselves mirrored in his painting as the genteel suitor who wins the pretty girl; the sportsman, bringing his rifle indoors and brandishing his powder horn as if just arrived from a turkey shoot, is, perhaps, a persona they left behind in previous circumstances. He is, in any case, what they do not wish to be. The expression on the sportsman's face conveys his dawning realization that his opponent has prevailed, yet he is without a clue why.

Guns in pictures of everyday life sometimes showed up hanging on the wall of a country home or tavern, such as in John Lewis Krimmel's *Barroom Dancing* of around 1820 (fig. 48). The arms were usually dusty and rusting, reminders of battles of yore and barnyard varmints rubbed out. *The Image Pedlar*, painted by Francis Edmonds around 1844, includes a rifle in the background of a humble kitchen where a family gathers around an itinerant vendor (fig. 49). Traveling salesmen were

FIG. 49

Francis William Edmonds
The Image Pedlar
about 1844, oil on canvas,
33 ¼ × 42 ¼ in. Collection of
The New-York Historical Society

familiar figures on American roadways, suspect individuals who criss-crossed the countryside to sell patent medicines, tinware, and notions to rural households. Some specialized in minor art objects and were self-taught artists who made painted and cut-paper portraits of children, uncles, aunts, and even family pets. Zeus-like, Edmond's beaming "pedlar" seems to hatch the statuette of Napoleon and other notables from his head while a grandfather and his grandson admire the vendor's bust of Washington on a table nearby. The child, dressed in a Continental army costume, listens to the old man tell of victories long ago; above them the trusty flintlock looms in the shadows. The great general's noble profile is set in the sunlight of an open window, while Napoleon's diminutive figure connects with the trigger end of the rifle on the wall. Washington and Bonaparte, liberty and tyranny—the image peddler is here the agent of both. Perhaps it is merely a coincidence that the procurer of images stands in the painting on the side of the tyrant, a symbol of the nefarious powers of art in the wrong hands. In any case, Edmonds himself is peddling a vision of a peaceful republican family in which political oppression is part of the past and the gun remains an ignored artifact.[86]

The American militia was an occasional subject in genre works and here, too, guns served mainly as window dressing rather than leading elements. Taking to heart the Second Amendment of the U.S. Constitution, which states that a "well-regulated Militia" was "necessary to the security of a free State," nineteenth-century citizen groups regularly rallied in local militias and held training sessions. The Fourth of July was a favorite day for militia drills, an event of wearing uniforms, carrying flags, firing guns, and singing patriotic songs. Commanders complained vociferously about volunteers' behavior on such occasions and their failure to muster with guns that worked—when they brought them at all. In *Militia Training*, a painting of 1841, James Goodwyn Clonney offered a wry commentary on the men who reported for militia duty (fig. 50). The scene shows no orderly assembly of troops or indication of who is in charge; indeed, the men of the militia unit are impossible to make out, for all is mayhem. In the foreground at the center of the canvas, a man in a partial uniform sways in front of a cart that hauls a whiskey barrel, and everyone present seems to have partaken in this ardent refreshment; even a barefooted boy is at the spigot to top off his cup. At the left of the image, a man in a beige jacket slumps with a hand to his forehead, opposite one of the fallen in the foreground who, too drunk to stand, crawls after his hat. The lone gun in the picture lies in the dust behind the prone man's sprawled legs. In the distance, the keen-eyed viewer can make out Clonney's final point about the United States' "well-regulated Militia." At the rear of the melee, the only man to muster in the feathered hat, red jacket, and white pants of full uniform is being ridden out on a rail by a group of young toughs.[87]

While firearms had a benign presence in works of art depicting average farmers and townsfolk, the diverse American audiences who appreciated those works were expressing concern about increasing gun

FIG. 50

James Goodwyn Clonney
Militia Training
1841, oil on canvas, 28 × 40 in.
Courtesy of the Pennsylvania Academy
of the Fine Arts, Philadelphia, Bequest
of Henry C. Carey (The Carey Collection)

violence in the United States. American political leaders were among
the first implicated in that trend. The election of Andrew Jackson in
1828 placed in the White House a man famous for his willingness to use
a pistol to defend his point of view. In 1803 he shot at the governor of
Tennessee, who had made a disparaging comment about Jackson's wife,
Rachel. "Great God!" thundered America's future president. "Do you
mention her sacred name?" Shots were exchanged, and one bystander
was grazed by a bullet. In 1806, after an argument over a horse race that
also included a criticism of his wife, Jackson challenged a Nashville man
to a duel. His opponent, Charles Dickenson, fired first, hitting Jackson
in the chest. Jackson shot and killed Dickenson, but he carried the
bullet in his chest until he died. Jackson shrugged off his wound. "If he

had shot me through the brain, sir, I should still have killed him."[88] "Old Hickory" sustained additional bullet wounds in a barroom brawl in 1813. Rather than detract from his reputation, however, Jackson's handiness with a gun enhanced his public reputation as the capable defender of the common man. By the time he took office, Americans had become inured to the common spectacle of public figures solving their disagreements with pistols.

Aside from the eminence of the two men, it was not a particularly unusual event when Vice President Aaron Burr killed the famed statesman Alexander Hamilton in a duel in 1804. After Hamilton publicly opposed Burr's candidacy for the New York gubernatorial election, Burr demanded satisfaction for the offense. Friends tried to intervene, but the men were resolute in their determination to have a confrontation. After Hamilton's death, the notorious precedent of their duel sparked a rash of congressional duels. The grassy meadow at Bladensburg field outside Washington, D.C., became the favored spot for the deadly encounters, and tourist groups soon arrived to see the place where more than one hundred politicians and other public figures had died. The vocabulary the men used to precipitate duels was simple and specific: the challenger needed to call his opponent a "liar,""coward," "rascal," "scoundrel," or, the unthinkable: "puppy." As an additional provocation, many of the insults were communicated publicly through newspaper interviews. Burr, Hamilton, and scores of other American leaders endorsed the code of honor that dueling represented. Senator Louis T. Wigfall of Texas said that dueling "engendered courtesy of speech and demeanor [and] had a most restraining tendency on the errant fancy, and as a preservative of the domestic relations was without equal." In 1841, English writer J. G. Milligan disagreed. "Duels in America [are] in general marked with a character of reckless ferocity...that clearly shows the very slow progress of civilization in that rising country."[89] The fiery clergyman and orator Lyman Beecher denounced the practice of dueling as a "great national sin."[90]

As pistol-packing legislators became models of armed violence, even ordinary people began to carry weapons. The problem of armed violence was more frequent in the borderlands and the south, but by the 1830s even people in the towns and cities of the east feared a violent breakdown of law and order. Rumors circulated about the subversive intrigues of Mormons, Catholics, Masons, land agents, and financiers; brawls in taverns and theaters, labor strikes, and riots became common events. The consumption of whiskey doubled. The editor of the *Niles Register* of Baltimore wrote in 1835, "We have executions and murders, and riots to the utmost limits of the union. The character of our countrymen seems suddenly changed, and thousands interpret the law in their own way ... guided apparently by their own will." Philadelphia's *Public Ledger* decried a recent spate of gun murders. "Every day exhibits some portion of the sovereign people in arms against the laws of God and the country, and against their own rights and the rights of others." The Boston *Evening Transcript* in 1841 declared that there had never been such an "extensive system of frauds, villainies, and robberies, and all kinds of rascalities."[91] Abraham Lincoln, in a lyceum address of 1838, spoke of "increasing disregard for law which pervades the country." He insisted that to take the law into one's own hands was "to trample on the blood of [one's] father, and to tear the [charter] of his own and his children's liberty."[92] Apparently, in their hunt for the social and economic opportunities promised by Democratic Time, many citizens were finding the gun to be an expeditious alternative to the tedious process of reaching an agreement or the even more disagreeable prospect of sharing.

The European immigrants who flocked to the United States in the 1830s and 1840s fueled American fears of social upheaval. Word circulated that officials in Ireland, England, and other European countries had arranged to ship off their paupers to the United States, and people on the lower rungs of American society railed about foreigners who were ready to steal their jobs for lower pay. People of a "better sort" complained about a lowering of standards. Writing in his diary in 1838, the New Yorker George Templeton Strong expressed the revulsion many Americans felt for the newcomers. "It was enough to turn a man's stomach—to make a man adjure republicanism forever—to see the way they were naturalizing this morning at the hall.... The very scum and dregs of human nature filled the ... office so completely that I was almost afraid of being poisoned by going in." He added: "A dirty Irishman is bad enough, but he's nothing compared to a nasty French or Italian loafer."[93]

With a flood of impoverished Europeans competing with the native born for jobs that paid less than a subsistence wage, reform groups, railroad companies, and manufacturers came together to lobby for opening the western territories—and new markets—to eastern wage earners. Large land companies had monopolized holdings in the West through the 1830s, but Congress passed new legislation in the 1840s that allowed those who migrated west to purchase public lands cheaply. Suddenly, droves of common people were loading their wagons and heading to the borderlands to take advantage of the opportunity to speculate on western real estate; most purchased a small parcel and lived there until they realized a profit, then cashed out and moved to a new plot. Every easterner who stayed home had a brother, an uncle, and several neighbors who went west. New Englanders, New Yorkers, and Pennsylvanians often set out for Ohio, while Virginians, Carolinians, and Georgians favored Kentucky and Tennessee. Texas also beckoned after the United States annexed it in 1845, as did the extensive regions ceded by Mexico in 1849 at the end of the Mexican War, including present-day Arizona, Nevada, New Mexico, and California. Adding the lure of mineral riches to the attractions of appreciating property, the California gold rush of 1849 drew hundreds of thousands of prospectors—and the grocers, launderers, and wagon makers who supplied them—to the mines of the Sierra. Families at home devoured letters from loved ones scattered across the West and visualized the strange and dangerous circumstances they faced in distant places. Those who went west wrote about feeling transformed by their experience and wondering who they had become.

Enter the frontiersman, the rough-and-ready, straight-shooting western trapper and scout who rode into artworks in urban galleries and magazine and book illustrations seen by millions of nineteenth-century Americans. With a stroke of the brush, the violence of eastern towns and cities seemed to retreat to an always-distant frontier that no one could specifically locate. Artists discovered that satisfying the public's growing appetite for pictures of buckskin-clad adventurers was refreshingly remunerative. The new American hero emerged on canvas as a man who knew his way around a rifle and, if the blending of uncouth frontier ways with the cultivated sensibilities of art was something of a contradiction, viewers embraced it as a fitting expression of the national character. The art critic Henry Tuckerman declared in 1846 in *Godey's Lady's Book* that "it is in our border life alone that we can find the materials for our national development as far as literature and art are concerned."[94] American audiences already knew frontier heroes from best-selling biographies about Daniel Boone and Davy Crockett and novels about western trappers and mountain men

was the "prairie cavalier" preparing to fight Indians, fighting Indians, and outsmarting and outshooting Indians. The superiority of white marksmanship was never in doubt in such works, and the happy ending of a crowd-pleaser like *The Trapper's Last Shot* was always understood.

Americans rooting for the frontiersman in images showing mortal combat with Indians stood square in the crosshairs of Samuel Colt, the owner of a spanking new firearms manufactory in Connecticut (fig. 53). After inventing an improved revolver that could fire six shots without reloading, Colt traveled around the country on the lyceum lecture circuit to raise money for his business. He assumed the persona of "Dr. Coult," a "practical chemist" from "New York, London, and Calcutta." Dr. Coult's specialty was laughing gas, and he charged fifty cents for his demonstrations of the hilarious discombobulating of spectators under the influence of nitrous oxide. His shows earned enough money to open his factory, and he gained valuable experience in not only altering his customers' state of consciousness but also making fools of skeptics. Through the aggressive promotional campaign that he waged until his death early in the Civil War, the Colt six-shooter became one of the most popular weapons in the United States and a firearm praised by Texas Rangers and fearful urbanites alike. Colt pressed his advantage in the art of stagecraft, presenting elegantly cased sets of pistols to congressmen and magistrates and blanketing public places with flyers and broadsides advertising his weapons. He also hired influential

FIG. 53

Unidentified artist
A Day at the Armory of Colt's Patent Fire Arms Manufacturing Company. From *United States Magazine* 4, no. 3 (March 1857)

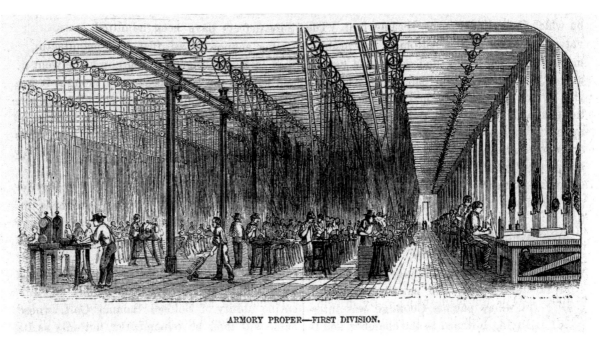

ARMORY PROPER—FIRST DIVISION.

people, including the well-known artist George Catlin, to give testimonials for his arms. Colt gave a simple reason for buying his products. "The good people of this world are very far from being satisfied with each other and my arms are the best peacemakers."[102]

By the 1850s, Colt's pistol was already the stuff of legend. Its appeal was enhanced by the growing fame of his factory in Hartford, the largest and most advanced armory in the world. Journals around the nation ran illustrated articles about its operations and, in the national imagination, the images of its astonishing technological advances melded with pictures of exhilarating frontier adventures. The Colt factory was a hall of wonders, an immense workshop filled with a vast and obedient army of jigging machines, gun-stocking machines, turret lathes, screw machines, barrel-boring machines, and drop forges (fig. 54). Visitors called it "magical," a "miracle of machinery," and a sight "almost beyond the power of the imagination to conceive."[103] After a tour of the establishment, even the famously irreverent Mark Twain gushed about its "tangled forest of rods, bars, pulleys, wheels, and all the imaginable and unimaginable forms of mechanism" where a visitor could "stumble over a bar of iron as he goes in at one end of the establishment, and find it transformed into a burnished, symmetrical, deadly [revolver] as he passes out at the other." Twain concluded, "it must have required more brains to invent all those things than would serve to stock fifty Senates like ours." The company made sure that the press paired admiring testimonials from factory visitors with stories about Colt's beneficence as employer. These often centered on Coltsville, the new town built next to the factory that the company filled with pleasant homes, churches, and music halls for workers' convenience and well-being. By midcentury the Colt revolver gleamed in the public mind as the icon of American ingenuity, compact and burnished evidence that the United States had come of age.

Nonetheless, Twain seemed finally to shake off the enchantment of his factory visit when he concluded his observations with a reminder of its dark purpose. "I took a living interest in that birth-place of six shooters, because I had seen so many graceful specimens of their performances in the deadfalls of Washoe and California."[104]

In the company's advertisements the Colt revolver became a passport to adventure (fig. 55). One of its broadsides published in the early 1850s unveiled the exciting places where customers could go and the rip-roaring things they could do when they held Colt weapons in their hands. Although most of Colt's customers were in the East, the

FIG. 55

Advertising broadside for the Colt's Patent Firearms Manufacturing Company, about 1854. Private collection

FIG. 56

George Catlin
Catlin the Artist Shooting Buffalos
with Colt's Revolving Pistol
1855, oil on canvas, 19 × 26 ½ in.
Wadsworth Atheneum Museum of Art,
The Ella Gallup Sumner and Mary
Catlin Sumner Collection Fund

company's publicity showed a firearm that belonged in the West. An enlargement of scenes engraved on the cylinders of high-end models, the advertisement includes images of an Indian raid, a seafaring skirmish in Texas, and a frontier holdup, resolving finally in a detailed design of a revolver. Samuel Colt also commissioned George Catlin to depict himself in action with a Colt as he explored faraway places. Always in debt and delighted to undertake a paid hunting excursion, Catlin obliged with a self-portrait that showed him "peacemaking" with buffalo in the remote reaches of the frontier; the work was reproduced as a lithograph that could be bought in black and white or in a colorful hand-tinted version (fig. 56).[105]

FIG. 57

Unidentified photographer
George W. Northrup Posed with
Gold Mining Equipment
about 1858, daguerreotype.
Minnesota Historical Society

Shooting the shaggy beasts with a handgun was more likely to annoy them than kill them, but the urban dwellers that comprised the main audience for Catlin's picture were not interested in ballistics. Buffeted by the winds of enterprise and fettered by the work disciplines of the nation's galloping industrialization, Colt's customers were interested in escape. The company's promotions beckoned with the promise of freedom, if only for their alter egos. George Northrup, a Midwestern schoolteacher preparing to embark for California after the peak years of the gold rush, stopped first at the studio of a daguerreotypist (fig. 57). A Colt .31 caliber "Baby Dragoon" is tucked neatly into his belt. The Colt had emerged as the preferred weapon of every California adventurer; the clamor for the guns was so great during the tumultuous years of the gold rush that even the "magical" factory was unable to keep pace with it. Evoking visions of hardships endured and bargains struck to secure one, the schoolteacher's Colt is a rare prize. For reasons unknown, the hopeful miner never departed for California, but the image remains as a record of his aspirations. On his lap, nestled next to his Baby Dragoon, is the sham happy ending of the prospector's fantasy: a bag—suspiciously fluffy—that reads: "GOLD $90,000."

For American arms makers, the Civil War was the mother lode of their dreams. Inconveniently, Colt died soon after the war began, but the directors of his factory joined scores of other American companies responding to the Union's urgent call for arms. Northern manufacturers produced some three million firearms during the four years of the war for a Union army estimated to have numbered 2,100,000 men.[106] The South was cut off from previous suppliers and settled for guns it could capture from the enemy and weaponry imported from Europe. The mechanical marvels and mass-production techniques pioneered at Colt's and other armories allowed the Union to churn out weapons in quantities never seen before. Northern improvements in weaponry, including metallic cartridges, rapid-fire magazine-fed guns, large-rifled cannon, "ironclad" warships, and rotating turrets for naval guns, also improved the killing capacity of the Union arsenal. Battling an enemy that commanded the most technical and scientific military effort to that date, the South suffered devastating casualties, not only because their technologies were inferior but also because they failed to adapt their tactics to the North's advanced gunnery. In 1862, as Confederate soldiers were being mowed down by quick-loading Union rifles, a Southern newspaper championed a time-honored plan of attack: "The greatest minds in the South are coming to the conclusion ... that our liberties are to be won by the bayonet, since no Federal regiment can withstand a bold and fearless bayonet charge."[107]

A picture of a Union rifleman that appeared in *Vanity Fair* in 1861 expressed the rumblings of unease that accompanied the Northern public's reception of newfangled killing machines. In the cartoon "The Patent Chronometer Telescope Rifle" by Elihu Vedder, a mustachioed recruit grapples with the whys and wherefores of his weapon (fig. 58). The toolbox at his side and the hammer at his feet hint at complicated assemblies of its interchangeable parts and the challenge for troops expected to rely on such contraptions in the heat of battle. A longstanding desideratum of American military leaders, a functional telescopic firearm was the focus of experiments carried out by Charles Willson Peale and the eminent Philadelphia scientist David Rittenhouse during the War of Independence. The two failed to devise a telescopic gun for Washington's troops, however, and colonials fell back on the expert aim of men from the backwoods.[108] In Vedder's illustration the legendary marksman of the Revolution becomes an operative trying to make sense of a collection of technological appurtenances.

Working as an "artist-correspondent" for *Harper's Weekly* during the Civil War, Winslow Homer took up the subject of the telescopic rifleman in a painting and in an engraving, published in the magazine in 1862 (fig. 59). In "The Army of the Potomac—Sharpshooter on Picket Duty," the humor of Vedder's befuddled gunner evaporates, replaced by Homer's hard look at the deadly work of a Union sniper. Splayed awkwardly in the branches of a pine, the sharpshooter balances himself with his left hand while his right holds his telescopic rifle. Such weapons were heavy—fifteen or twenty pounds—and required their handlers' considerable strength to hold them in position while searching for a target and waiting to squeeze off a shot. Set off against the glint of gunmetal, the soldier's index finger poised on the trigger is curiously delicate. Gentle contact was essential, since sharpshooters' guns were fitted with hair triggers that fired with the pressure of a caress. One gunning expert spoke of "that delicate sympathy, which ought always to exist between hand and eye."[109] The "delicate sympathy" of the soldier's trigger finger connects to the life of an unseen Confederate that hangs in the balance as he makes his selection through his telescopic sight. Aloft in the

FIG. 58

Elihu Vedder
Camp Sketches: The Patent
Chronometer Telescope Rifle.
From *Vanity Fair*, November 16, 1861

THE PATENT CHRONOMETER TELESCOPE RIFLE.

FIG. 59

Winslow Homer
***The Army of the Potomac—A
Sharpshooter on Picket Duty***
1862, wood engraving on paper,
9 ⅛ × 13 ¾ in.
National Museum of American
History, Smithsonian Institution

boughs of a pine—the type of tree that furnished wood for common coffins—the gunner is an angel of death. Homer declined to ennoble his business, producing a reportorial account that seems to leave his own judgments and emotions out of the picture.

Sharpshooters were called for duty when there was a lull in the action and exhausted soldiers from both North and South were resting. Reviled as "murderers" by both sides, they usually were summarily executed if captured. An infantryman described the terrible suddenness of a sharpshooter attack in a letter to his family from the front in Virginia. "There was a man this moment shot in the belly twenty feet from me which is nothing unusual in this country."[110] Years later, in 1896, Homer himself sketched and recounted his experience peering through the scope of a telescopic rifle in a company of Union sharpshooters (fig. 60). "I looked through their rifles once when they were in a peach

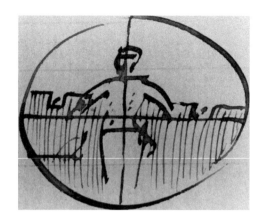

FIG. 60

Winslow Homer
View through telescopic rifle, 1896,
detail from ink drawing on paper
(letter to George G. Briggs). Winslow
Homer Papers, Archives of American
Art, Smithsonian Institution

orchard in front of Yorktown in April, 1862," he remembered. "This is
what I saw—I was not a soldier—but a camp follower and artist. The
above impression struck me as being as near murder as anything I ever
think of in connection with the Army and I always had a horror of that
branch of the service."[111]

Despite its apparent detachment, "Sharpshooter on Picket Duty"
betrays the artist's revulsion for his subject with the shooter's dangling
buttocks, the eager inclination of his body, and the humanity of his face
blotted out by a technological innovation. Hanging from the boughs like
a malevolent monkey, the soldier introduces the strange Liberty Tree of
modern warfare. The picture concedes nothing to one of the famous few
who were Lincoln's darlings during the war. The soldier on picket duty
belonged to Colonel Hiram Berdan's celebrated New York Company
of Sharpshooters, named after their weapons, the Sharps model 1859
breech-loading open-sight rifle. Lincoln not only took a keen personal interest in the Sharpshooters' training and assignments but
also made sure they were equipped with the latest weaponry. Taking
time out of his punishing wartime schedule, the president visited the
Sharpshooters' camp in 1861 to watch the men at target practice. They
invited Lincoln into one of the rifle pits, where he squeezed off three
respectable shots. "Boys, this reminds me of old-time shooting," he
declared as a cheer went up from the Sharpshooters. His generals had
advised him to use time-tested frontloading guns, but Lincoln himself
gave orders to supply Berdan's sharpshooters with breech-loading
rifles, hair triggers, and telescopic sights. When composing his picture,
Homer ignored the aura of glamour that surrounded Berdan's gunners.
Instead, he made the readers of *Harper's Weekly* witnesses to a cold-
blooded ambush, a summary execution without courage or honor.[112]

When a machine gun was first used on a Virginia battlefield in 1862,
it appeared that the entire question of marksmanship would soon be
moot. The "coffee mill" gun, named for a crank at the back that was
turned to churn out bullets, allowed the operator to point and spray
without worrying about accurate aim. Although the weapons' tendency
to overheat and jam made them a disappointment during the Civil War,
gunmakers continued to promote them aggressively. The Gatling gun,
in particular, was backed by a broad-reaching advertising campaign
that brought it to the attention of both the military and the general
public (figs. 61 and 62). Addressed to Union commanders, one of the

FIG. 61

F. L. Seitz
The Gatling Gun.
From Horace Greeley et al.,
*The Great Industries of the
United States* (Hartford, CT:
J. B. Burr and Hyde, 1872)

FIG. 62

Unidentified artist
Cartridges Used in the
Gatling Gun.
From Horace Greeley et al.,
*The Great Industries of the
United States*

company's promotions pointed out the efficiencies of the novel weapon. The Gatling gun would "more than double the strength of our armies in the field" and "lessen the number of men required, thereby removing many of the difficulties attending the procuring of men by draft, etc." It would also "save lives, wounds and sickness, by lessening the number subjected to the perils of war" and lead to great savings by "lessening the amount to be paid for bounties and pensions, arming, clothing, feeding, and transporting the troops." The advertisement concluded by dispensing the advice that the federal government needed to avail itself "of all the advantages, which a kind Providence has placed within its reach to crush the rebellion."[113] In short, the Gatling gun was a time-saver and, therefore, a money-saver that applied the efficiencies of peaceful production to the business of warfare.

It is estimated that 620,000 men died in the Civil War.[114] From the country's population of men of military age, more than one in four perished, most of them farmers or farm workers before the war. Although French and Fancher's ordnance plow (see fig. 46) was never produced, the harvest of death by the new and old gunning machines was beyond what anyone had dreamed at the onset of hostilities. Firearms cannot be awarded full credit for the final body count—disease and infection were the most efficient killers—but they were the cogs, cams, and levers of a brutal military machine that Americans fueled with their hatreds.

THE GATLING GUN.

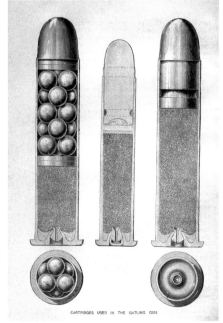

CARTRIDGES USED IN THE GATLING GUN.

These passions did not subside after the surrender at Appomattox; many men departed the battlefield with a burning desire for revenge. Most also left well equipped. The war created arms that were consistently reliable, trained millions to use them, and gave soldiers from both North and South weapons to bring home. Civilians who earlier had been unable to afford the expense of a decent weapon could now make a selection from the vast stockpile of surplus arms that the federal government put on the market at bargain prices. The number of American arms makers mushroomed during the war, with the yearly production of small arms expanding from thirty thousand at the time of the attack on Fort Sumter to more than seven hundred thousand by the start of Reconstruction.[115] The Civil War not only affirmed the political union of the United States; it also endowed the country with the most advanced weaponry in the world.

As units mustered out to return to their families and former occupations (if they still existed) and a much-changed nation, game hunters began to trickle into works of art and popular imagery. By the 1870s and 1880s a veritable stampede of "sportsmen" were stalking their prey

FIG. 63

John Whetten Ehninger
Turkey Shoot
1879, oil on canvas, 25 × 43 ⅜ in.
Museum of Fine Arts, Boston, Gift of Maxim Karolik for the M. and M. Karolik Collection of American Paintings, 1815–1865

FIG. 64

Unidentified artist
Martha Maxwell, "The Colorado Huntress"
about 1870, gelatin silver print, 12 × 18 in.
Courtesy, History Colorado

FIG. 65 ☞

Martin Johnson Heade
Marsh with Hunter
1874, oil on canvas,
15 × 30 ⅛ in.
Private collection

on canvases by famous artists; in Currier and Ives prints and photographs; and peeking out from underneath the mashed potatoes on dinner plates and firing away on decorative wallpaper, music sheets, and whiskey-bottle labels. Americans from every region and every rung of the social ladder made an appearance in such images—New England Brahmins, "middling" folk from the Midwest, Southern freedmen, and, from the Rockies, even refined ladies (figs. 63 and 64). Elisha Lewis, the author of a popular hunting manual, expressed the egalitarian spirit of such pictures when he invited "one and all" to join him in the field:

Is there not a time when the wan-faced student of science may neglect for a while the sickly flickerings of the midnight lamp? Is there not a time when the learned counselor may escape the wranglings, the jeerings, the bitter feuds of the halls of justice? And is there not a moment of leisure, an hour of repose, when the skilful [sic] physician may turn a deaf ear to the harassing solicitations of suffering humanity, and draw for a brief period the curtain of oblivion around the couch of disease and death?...Yes! There is a time, thanks to the noble founders of our liberal institutions, when even the industrious artisan, freed from all care and anxiety, may forget the labors and duties of the shop, and wander forth to enjoy the wonders of nature and learn to appreciate the boon of freedom, his country's dearest gift. To the fields, then,—to the bright and beautiful fields,—with "dog and gun," do we invite you, one and all, to spend those hours of leisure and participate in those innocent enjoyments so captivating to a true sportsman.[116]

Pictures of bright and beautiful fields showed the possibility of a true national re-union, an old-time hunters' rendezvous that would weave back together a country that had been torn asunder by war. On canvas and in giftbooks and illustrated outdoor guides, Americans from every station were seen tramping in the woods. Shooting manuals became best sellers; they described the hunt as a sport of patience, accurate observation, and physical strength. A return to bygone values and simpler times, the hunt was also a project of self-government in which sportsmen trained themselves to conquer their fears and excitements. "Success in shooting . . . is dependent in a great measure upon coolness and deliberation," Elisha Lewis explained. "Mastery over [the hunter's] feelings being once acquired, there will be no difficulty in the way of rapid progress."[117]

Marsh with Hunter, painted in 1874 by the artist and avid hunter Martin Johnson Heade, envisioned one of his favorite pastimes: wing shooting, the sport of shooting birds in flight (fig. 65). The most popular

type of hunting in the period, wing shooting took considerable time to master and required agility, stealth, and the ability to make quick decisions, in addition to impeccable aim. In the lower right corner of Heade's painting, a khaki-clad hunter behind a stand of grass looks out across the landscape, and birds strewn on the ground behind identify him as a respectable wing shot. Although the hunter seems to be resting, he is listening to the calls of a flock approaching in the distance, holding his rifle ready on his knee for the moment when they come into range. The sportsman is small, immersed in his surroundings, making Heade's point that appreciation for nature went hand in hand with hunting.

The artist located his hunting scene in the dawn mists of an Atlantic salt marsh, the open wetlands that were a favorite haunt of hunters from urban areas. These boggy places, an hour's distance from Boston or New York, were unsuitable for the construction of homes or shops but ideal for a salaried man's daylong hunting excursion. *Marsh with Hunter* conveys the solitary pleasures of a morning in the marsh, where quick-flying waterfowl—canvasback, curlew, woodcock, geese, and snipe—replace the crowds and the commotion of city life. The marshland and haystacks in the painting soothe the eye with their soft anonymity; although the details of grass, sunlight, and water are meticulously rendered, the work offers no specific features or landmarks that might allow a viewer to identify the place exactly. The painting asks the observer to attend patiently, as a good sportsman would, to its flat stillness. It is an exercise in the close-looking, time-taking virtues of the expert wing shot, as well as the camouflaged composition of an artist who had no intention of giving away the location of his favorite birding spots.

Heade painted *Marsh with Hunter* as Americans witnessed the last of the great buffalo herds passing into memory, wiped out by hide hunters who were shooting hundreds of buffalo on the prairie each day. Heade's hunting idyll was also set against the backdrop of a sudden public awareness of the precipitous decline in the number of birds in backyards, local parks, and forests and the circulation of popular prints that showed hunters ready to blast their quarry into oblivion (fig. 66). Beaver, fox, and other fur animals had long ago been decimated by indiscriminate "takes," but, while the artist was putting the final touches on his marsh picture, citizens were raising their voices against the senseless exterminations that were occurring all around them. Heade was one of many hunters calling fellow sportsmen to account. Writing under the pen name "Didymus"

FIG. 66

Currier and Ives, publishers
The First Bird of the Season
1879, lithograph, 9 × 13 in.
Museum of the City of New York,
The Harry T. Peters Collection

THE FIRST BIRD OF THE SEASON.

as a longtime contributor to *Forest and Stream* ("A Journal of Outdoor Life, Travel, Nature Study, Shooting"), he pleaded with his readers to honor the game laws that gave wildlife stocks a chance to regenerate.

In his writings for *Forest and Stream*, Heade held fast to the idea of a sportsman's code, the notion that those who hunted birds and other animals assumed an obligation of stewardship for the health of their populations. Heade admonished readers who took the woodcock, "the true sportsman's bird," out of season during its fledging time in the summer. The birds' numbers were dwindling to an alarming degree, he cautioned; he told of a recent excursion when he had seen hardly any of this once-plentiful quarry, warning, "If summer shooting is not prohibited, we will soon find the woodcock 'a scarce commodity.' "[118] The artist joined the growing ranks of late-century sportsmen who saw themselves as champions for the conservation of wildlife. Despite hunters' complicity in the indiscriminate slaughter of animals, their outings to the wilderness were forcing them to confront the fact that the bounty of days gone by no longer existed. In a curious and hopeful behavior reversal, many sportsmen's clubs and shooting societies around the nation transformed into powerful preservation advocates, marshalling their considerable resources and influence to bring several species back from the brink. *Marsh with Hunter*, with its little man set against the ample horizon of a foggy meadow, envisioned an equilibrium of humankind and nature. "The pleasure of . . . shooting does not consist alone in the mere act of killing a bird," Didymus insisted. "The soul of the true sportsman in also keenly alive to the beauties of nature."[119]

Still-life artist William Harnett had already created a number of paintings of hunting gear hanging on weathered doors when he executed a similar picture of a hanging pistol, *The Faithful Colt*, in 1890 (fig. 67). His sportsman-themed works depicted rifles, game, powder horns, flasks, and other hunting paraphernalia suspended between the elegantly wrought hinges of rustic doors. The artist gave an unsettling twist to his sportsman metaphor when he replaced the rifle, which his viewers understood as an arm for hunting or shooting contests, with a pistol—a weapon of interpersonal violence. Harnett's audience saw the pistol as the weapon of the indignant senator and the avenging husband, the rough-and-ready miner, and the back-alley hoodlum. Pistols were what ladies tucked in their purses before they embarked on railroad journeys and what desperados pointed at bank tellers. With such a wide range of users to choose from, Harnett refrained from including clothing or equipment that might narrow audiences' choices as they imagined the identity of the "hunter" in absentia. The artist barred everything extraneous from his image, creating a painterly showdown between the viewer and the solitary Colt of the title.

Harnett selected as his subject the most storied and sought-after weapon of the nineteenth century. Thanks to Samuel Colt's highly effective promotional campaigns—the slip of paper on the board is one of the company's advertisements—the Colt signaled to the painting's beholders not ignoble murder but adventure, independence, and justice. The gun's rusty barrel and cracking ivory handle mark it as the veteran of many campaigns and, hanging from its nail like a crooked question mark, it invites the viewer to ponder where it has traveled and what wrongs it has righted. The Colt is a proud reminder of America's famed ingenuity and technological innovations that by the 1890s had propelled the United States to the front of the world's industrial order. Moreover, the exacting insights of the painting's trompe l'oeil technique, the "fool the eye" realism of a master artist, embody the keenness of vision that nineteenth-century people believed was a special American talent.

The painterly sleight of hand of *The Faithful Colt* makes a fitting connection with old Dr. Coult's entrepreneurial talents, for it was the

FIG. 67

William Michael Harnett
The Faithful Colt
1890, oil on canvas, 22 ½ × 18 ½ in.
Wadsworth Atheneum Museum
of Art, The Ella Gallup and Mary
Catlin Sumner Collection Fund

money that Colt earned in the laughing-gas trade that allowed him to put his gun factory into operation. The painting allies illusionism and a pistol in the kind of alluring image that is deeply implicated in setting the stage for armed encounters between Americans over the course of the nineteenth century. The faithfulness in Harnett's title refers not only to the reliability of Colt's patented revolver technology but also to the idea that the weapon was a dependable solution to conflicts, a "peacemaker." In a book about the great industries of the United States, the editor Horace Greeley expounded on the nineteenth-century's hypothesis that the better the weapon, the better the peacemaking. "The ignorance and stupidity of the masses of men, played upon by the crafty few in every age, render it possible if not probable that eventual 'peace throughout all the earth' will come to the race only when the genius of invention shall have so thoroughly armed nations and individuals with most destructive weapons as to equalize their power, or render death a sure result to all combatants in the field of war," Greeley surmised. "As a step toward the achievement of so desirable a result, the vast progress in the improvement and speedy means of manufacture of firearms in the United States . . . should be a matter of pleasure and pride to every humanitary [*sic*] American."[120]

The countless nineteenth-century pictures of frontier sure shots, armed Indian marauders, and righteous military battles supported Greeley's position. If the work of democracy is consensus building, however, the search for a common ground of understanding is a citizen's most important charge. As the phenomenal success of the Colt Armory attests, the time-honored business of encouraging Americans' disagreements and obscuring their shared values is a lucrative enterprise. *The Faithful Colt*, like Colt's advertisements, is a tour de force of inventive powers of persuasion, but the painting ultimately falls victim to its own clever construction. By leaving the owner of the revolver out of the picture—an artful tactic that frees its audience to imagine all sorts of swashbuckling fellows—*The Faithful Colt* also prompts the sharp-eyed viewer to ask the obvious question: faithful to whom?

I remember my first and succeeding impressions of Niagara; but never did I see an incarnation of vast multitudes, or resistless force, which impressed me like the main herd of buffalo. The desire to shoot, kill, and capture utterly passed away. I only wished to look till I could realize or find some speech for the greatness of nature that silenced me.

Fitz Hugh Ludlow, *The Heart of the Continent* (1870)

The Buffalo

N ENTRANCE TICKET FOR CHARLES WILLSON PEALE'S Philadelphia museum featured a menagerie of curious animals under a banner headed "The Birds & Beasts will teach thee!"(fig. 68). Inspired by a passage from scripture, the greeting invited Peale's visitors to observe carefully the stuffed birds, squirrels, and sheep in his galleries. During his frequent evening lectures, Peale reminded mechanics, bakers, and other ordinary folk that, while divine authority had endowed humans with dominion over the earth's wild creatures, such power was accompanied by the obligations of stewardship. Studying the habits of wild creatures, he advised, was a way for democratic citizens to learn about wise governance from the natural laws that animals modeled.

FIG. 68

Charles Willson Peale
Entrance Ticket to the Peale Museum
1788, etching and engraving on cream
wove paper, 2 $\frac{11}{16}$ × 3 $\frac{3}{4}$ in.
Worcester Art Museum, Richard A. Heald
Fund and the Thomas Hovey Gage Fund

When ticket buyers first entered Peale's renowned galleries, they were confronted by a large stuffed bison that did not look happy to see them. A visitor in 1826 stated, "The physiognomy of the bison is menacing and ferocious, and no one can see this formidable animal in his native wilds, for the first time, without feeling inclined to attend immediately to his personal safety."[121] That was the first tutorial of the "Beast;" over the course of the century millions of bison would lead the people of the United States in an extended course of instruction. What the animals taught was so monumental and profound that, even today, Americans are still trying to comprehend its full meaning.

The bison's impressive size—eight to ten feet long and four feet wide between its horn tips—made it one of the largest animals in a land famous for its prodigious creatures. Its great dimensions also made it a strategic weapon in the United States' cultural skirmishes with Europe. At the time of the American Revolution, zoology emerged as part of a scientific battleground in which native-born naturalists fought the "dictatorial powers" of Old World science. Americans had territorial feelings about everything in their country, including its endemic diseases. "We true Americans," wrote the editor of the *New England Journal of Medicine and Surgery* in 1813, "consider the subject [of yellow fever] as our national property, and we are jealous of any attempts among Europeans, especially Englishmen, to assume any control over it."[122]

Americans were incensed when European naturalists described American flora and fauna as puny and degenerate versions of Old World specimens. Their chief spokesman was the notorious Count of Buffon, who blamed the reduced state of America's creatures on the land's damp soil and cold climate. The one animal that thrived in North America, he announced, was the pig. According to his investigations, more than any other animal, the "hog has thriven the best and most universally [there]."[123]

As Europeans tut-tutted about the diminished size of American wildlife, the American scientific community enlisted the bison as a hooved ambassador for everything that was large and magnificent about North America. European visitors asking pointed questions about the health of the United States' democratic "experiment" were directed to observe the great dimensions and countless numbers of bison as proof of the natural robustness that nurtured the nation's strenuous project of self-government. Few Americans, however, had actually seen a live buffalo in the first half of the century. Tales circulated about the great herds that had once populated the eastern regions of the country. Old-timers in Kentucky and Pennsylvania spoke of seeing buffalo in places that were now thriving towns, and they knew areas in surrounding woods where old buffalo trails could be found. Three or four feet wide, with earth tramped smooth and hard over ages of use, the trails were reminders of the wooly giants who once roamed in the land.[124]

Easterners reflecting upon the disappearance of bison from their states were adjusting to recent news, disseminated in geology texts, popular journals, and lyceum lectures, that species in nature could go extinct. Fossil and geological discoveries in Europe and America proved that the earth was not static or stable and had experienced waves of catastrophic changes that had destroyed successive populations over eons. The American public also learned from scientific reports that North America's environment, with its varied geography and climate, was particularly conducive to the rapid waxing and waning of species. Usually dismissed by Europeans as a place of peripheral scientific interest, the United States and its adjacent territories suddenly earned a reputation as a living laboratory where the forces of extinction could be witnessed in action. Contemporary authorities on the disappearance of species including the Scottish geologist Charles Lyell defined it as a salubrious winnowing process leading from the less advanced to the more advanced. They advised that the extirpations represented not lamentable events, but rather a welcome for ever-more-perfect forms. As citizens of a nation founded on principles of human betterment, Americans read their books with keen interest, proud to be at the forefront of human progress.

In 1830, the year Lyell published the first of his three-volume *Principles of Geology*, the artist-explorer George Catlin embarked on one of several journeys inspired by the idea that America's original inhabitants would soon melt away before the advance of civilization. Esteeming the value of their cultures, he made numerous paintings of

FIG. 69

George Catlin
Buffalo Bull, Grazing on the Prairie
1832–33, oil on canvas, 24 × 29 in.
Smithsonian American Art Museum,
Gift of Mrs. Joseph Harrison, Jr.

buffalo as part of his survey of Native American life. Catlin's pictures and their reproductions in his books helped acquaint eastern audiences with the continent's zoological plentitude. In *Buffalo Bull, Grazing on the Prairie* of 1832–33, Catlin singled out a solitary animal from the herds of local legend and geography books (fig. 69). The artist described his subject in his published narrative as a "noble animal . . . that roams over the vast prairies, from the borders of Mexico on the south, to Hudson's Bay on the north." He also warned that it was "one of the most formidable and frightening looking animals in the world when excited to resistance."[125] The painting referees a polite encounter, showing the animal as it paused, curious, to gaze back at the artist. Catlin usually preferred to portray the buffalo in action—grazing, wallowing, fighting, stampeding; in this image he allowed the animal to take the artist's measure. Its glistening black eyes are wide open, straining for a good look and asking, who are you? Bold and inquisitive, the being looking back at us is a member of a herd still accustomed to spending much of its time undisturbed by human predators.[126]

Catlin's frontier paintings authenticated explorers' reports of the buffalo's considerable dimensions and also verified their accounts of herds large enough to stagger the imagination (fig. 70). Stories about the massive herds started to circulate at the beginning of the century when Lewis and Clark claimed to have seen one that they estimated to number twenty thousand animals. A member of Stephen Long's army survey in 1819–20 backed up the explorers' claims, describing "immense herds of bisons [*sic*], grazing in undisturbed possession and obscuring with the density of their numbers the verdant plain; to the right and left as far as the eye was permitted to rove, the crowd hardly seemed to diminish, and it would be no exaggeration to say that at least ten

FIG. 70

George Catlin
Buffalo Herds Crossing the Upper Missouri
1832, oil on canvas, 11 ¼ × 14 ⅜ in.
Smithsonian American Art Museum,
Gift of Mrs. Joseph Harrison, Jr.

thousand burst on our sight here in an instant."[127] While traveling on the Santa Fe Trail in 1839, the frontiersman Thomas Farnham reported seeing bison cover the whole country for three days. He calculated that the herd covered 1,350 square miles.

Curiously, reports about apparently limitless numbers of bison were matched by a parallel stream of commentary in scientific journals, political speeches, and newspaper editorials that predicted the animals' imminent extinction. The sources reiterated the same message until easterners could repeat it in their sleep: the herds and the native peoples who hunted them must be ready to step aside to make way for progress. Cued by such expansive discourse, Catlin positioned himself to be the advance guard of the coming settlement. He demonstrated that he was abreast of the most recent intelligence when he forecast in 1832 that the bison was "so rapidly wasting from the world, that its species must soon be extinguished." He proposed what he called his "Indian Gallery" as a means of preserving on canvas the animals and the Native American ways of life that would soon exist only in memory.[128]

In *Buffalo Hunt under the Wolf-skin Mask*, also 1832–33, Catlin documented the ingenious subterfuge that Indian hunters used to approach the herds and their extensive knowledge of the abundant resource that sustained them with food and necessary materials (fig. 71). After watching their camouflaged forays into the middle of a dense group of animals, Catlin wrote that the buffalo "would stand unwittingly and behold [the hunter], unsuspected under the skin of a white wolf, insinuating himself and his fatal weapons into close company, when they are peacefully grazing on the level prairies, and shot down before they are aware of the danger."[129] Unlike weaponry or the speed afforded by a mount, wolfskin hunting rewarded the hunters who made the most accurate observations of the animals' habits and relationships. Letting the animals teach them, the tribesmen saw that adult bison would, at certain times, tolerate scavenging wolves in their midst. Catlin captured in his painting a dramatic performance by people whose survival depended on a nuanced understanding of their prey and the ability to accurately simulate the movements and attitudes of a competing predator.

Buffalo Hunt under the Wolf-skin Mask documents a hunting method that indigenous peoples on the plains had used extensively before the arrival of Europeans. Until the late seventeenth century, Midwestern tribes hunted and fished for a variety of animals, planted crops, and collected

FIG. 71

George Catlin
Buffalo Hunt under the
Wolf-skin Mask
1832–33, oil on canvas, 24 × 29 in.
Smithsonian American Art Museum,
Gift of Mrs. Joseph Harrison, Jr.

wild plants in different habitats during the cycle of the seasons.
Hunting bison on foot was part of a gathering circuit that included an
assortment of animals and plants—bighorn sheep, ground squirrels,
catfish, corn, squash, acorns, pine nuts, berries, and roots. When the
horse arrived on the plains in the late seventeenth century, it was soon
followed by European fur traders offering metal tools, cloth, and food-
stuffs in exchange for buffalo robes. Indians became reliable suppliers
for the emerging North American fur business and, responding to the

FIG. 72

John Mix Stanley
Buffalo Hunt on the
Southwestern Prairies
1845, oil on canvas, 40 ½ × 60 ¾ in.
Smithsonian American Art Museum,
Gift of the Misses Henry

insatiable demand for buffalo robes, began to specialize in bison hunting. The horse gave them the mobility and power to kill large numbers of bison for exchange, as well as the option to abandon the varied kinds of harvesting that had sustained them for thousands of years.[130]

Catlin promoted his paintings of Indians hunting buffalo on horseback as a glimpse of untouched Native American customs, but these images actually represented communities adapting to the presence of Europeans. A picture of a brave galloping after a buffalo was a portrait of a society meeting head on the changes that had come; eastern audiences, however, saw such images as evidence of cultures frozen in time and, thus, utterly different from their own. In his efforts to create authentic images of buffalo hunts during his long sojourn among the tribes, Catlin himself became an experienced hunter, pursuing his quarry with his drawing pencil as well as a gun. He recounted making a sketch of a dying buffalo that had been shot by an Indian's arrow. "[I] drew from my pocket my sketchbook, laid my gun across my lap, and commenced taking his likeness. . . . I rode around him and sketched him in numerous attitudes." When the wounded animal stopped struggling, he wrote, "I would then sketch him; then throwing my cap at him, and rousing him on his legs, rally a new expression, and sketch him again."[131]

By midcentury, pictures of buffalos and buffalo hunts were ubiquitous subjects in art galleries, school primers, and broadsides (figs. 72 and 73). Many of the artists producing the scenes had never visited the West, but they enthusiastically pieced together elements from works by other artists and their own imaginations. A galloping bison borrowed from Catlin might be paired with one of John Mix Stanley's Indian braves and rounded out with a section purloined from a French master. In these assemblages the accurate ethnographic, topographical, botanical, and zoological details that distinguished portrayals by Catlin and other explorer-artists dropped away, replaced by good storytelling. Eastern audiences especially enjoyed seeing pictures of American frontiersmen on buffalo hunts. For Europeans, big game hunting was a closely guarded privilege, and poachers faced steep fines, imprisonment, or death. A picture of a common fellow taking aim at a prize specimen enacted a bit of nose thumbing at the pedigrees of the Old World and filled citizens with patriotic pride. Furthermore, while the extent of Europe's forests had shrunk to a fraction of its former size and the number of its large mammals had declined precipitously during the previous centuries, North America's extravagant numbers of wild animals guaranteed that any able-bodied person could enjoy the pleasures of the chase.

Despite the egalitarian availability of buffalo, most easterners could not afford the expense of a journey to the plains or the time it took to travel there and back. P. T. Barnum, the New York impresario who made a profitable

FIG. 73

Felix Octavius Carr Darley
Buffalo Hunt.
From John Frost, ed., *Book of the Indians of North America: Illustrating Their Manners, Customs, and Present State* (Hartford, CT: W. J. Hamersley, 1852)

business out of understanding the deepest longings of the American common man, intuited in the 1850s that every red-blooded resident of New York City yearned to test his mettle in a buffalo hunt. Barnum knew that women-folk and children wanted to witness the mettle testing, too. Happy to oblige, Barnum invited the people of New York to a "Grand Buffalo Hunt," free of charge, to be held in Hoboken, New Jersey. He told the prospective hunters that they needed only to cover the cost of their ferry tickets across the river, after he had arranged, of course, to rake off a substantial portion of the sales. The herd was ready: fifteen "remarkably docile" yearlings that had been driven recently to New Jersey from the plains. Barnum complained in his autobiography that the "poor creatures were so weak and tame that it was doubtful whether they should run at all, notwithstanding my man French had been cramming them with oats to get a little extra life into them." On the appointed day, having paid the steep ferry passage, twenty-four thousand hunters showed up, raring to go. But despite the handlers' best efforts, the calves refused to move and cowered against a fence. Expressions of anger about the humbug rippled through the throng and eventually turned to peals of laughter. The noise spooked the little bison and they began to run. Managing to escape their corral, they scampered off to a nearby swamp as the crowd stampeded in fear.[132]

Buffalo hunting became big business by the 1870s, spurred by the enormous expansion in American manufacturing and transportation following the Civil War. A buffalo weathervane, one of the mass-produced consumer items advertised by the Mott Iron Works as "first class work at moderate prices," is an emblem of both the economic surge and the gathering storm that would sweep away the last remaining herds in the West (fig. 74). Technological changes taking place far away

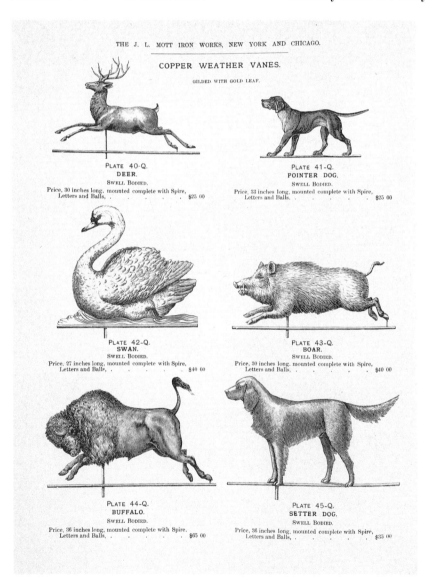

FIG. 75

After James Henry Moser
A Still Hunt.
From William T. Hornaday, *The Extermination of the American Bison* (Washington, D.C.: Government Printing Office, 1889)

from the grasslands of Kansas and Illinois would prove to have a lethal effect on the prairie. The military's demand for more reliable guns during the Civil War led to easier-to-load and more accurate rifles; the conflict also produced armies of men adept at using them. Thanks to improved tanning techniques, buffalo hides were adopted for industrial belting in factories that were rapidly expanding production of farm implements, industrial machinery, and consumer products. Furthermore, the completion of the transcontinental railroad in 1869 made it easy for rifles and hunters to go west and buffalo hides to go east, while the introduction of the refrigerated railroad car meant that meat from cattle raised on western ranches—land where buffalo had ranged—could be shipped to eastern markets. The low cost of prairie rangeland and corn made western beef 5 to 10 percent cheaper than the local product, and customers all over New England lined up to buy it. Cattle ranching in the West, once a hardscrabble business for people without better options, became a viable and respectable enterprise.[133]

By 1872 thousands of hide hunters arrived on the prairie, many of them toting the Sharp Rifle Company's new "big fifty" (fig. 75). The gun fired huge slugs, each weighing up to one pound and containing ninety grains of gunpowder; the deadly weapon allowed hunters to drop a buffalo from several hundred yards away. During the hide boom from the early 1870s to the early 1880s, reasonably industrious hide hunters killed 150 to 200 or more buffalo a week, earning ten to twenty dollars per day at a time when western sheriffs earned $150 to $250 a month. The only workers of that period who earned more than hide hunters were prostitutes.[134]

Beginning in 1869, railroad companies offered bargain fares for daylong buffalo hunts from Chicago, Cincinnati, St. Louis, and other cities adjacent to the prairie. During these jaunts passengers were allowed to fire their weapons from the windows and tops of trains—thus

THE FAR WEST.—SHOOTING BUFFALO ON THE LINE OF THE KANSAS-PACIFIC RAILROAD.

FIG. 76

Ernest Griset
The Far West.—Shooting Buffalo on
the Line of the Kansas-Pacific Railroad.
From *Frank Leslie's Illustrated
Newspaper*, June 3, 1871

experiencing a taste of authentic frontier adventure without the rigors and expense of a traditional hunt (fig. 76). The artist Theodore Davis, who joined a railroad hunting excursion as a reporter for *Harper's Weekly*, described the action. "It would seem to be hardly possible to imagine a more novel sight than a small band of buffalo loping along within a few hundred feet of a railroad train in rapid motion, while the passengers are engaged in shooting, from every available window, with rifles, carbines, and revolvers," he wrote. "An American scene, certainly." Since parlor cars on western lines often featured pianists and musical entertainments, passengers may have heard the sound of the pianoforte against a background ruckus of rifle shots, clacking of wheels on track, and the groans of expiring buffalo. Perhaps the stirring company anthem, the "Kansas Pacific Grand March," wafted across the prairie, attuning their passage through buffalo land with the steady march of progress (fig. 77).[135]

Among the many nineteenth-century Americans who protested vehemently against the slaughter of the bison, George Catlin was one of the earliest to speak up. Despondent about the scale of the killing, he was also one of the first to widely publicize a solution for the buffalo. Envisioning the creation of a vast national preserve, Catlin recommended that the federal government create "a nation's Park" in the grasslands where both native peoples and the bison could live undisturbed. In *Notes* he appealed to his readers to share in his vision. "And what a splendid contemplation, too, when one (who has traveled these realms, and can duly appreciate them) imagines them as they *might* in future be seen (by some great protecting policy of government) preserved in their pristine beauty and wilderness, in a *magnificent park*."[136]

FIG. 77

Kunkel Brothers, publishers
Cover for sheet music, "Kansas Pacific
Grand March," 1872

Catlin's entreaties were ignored, but the number of American voices calling out in protest swelled to hundreds of thousands over the following decades. In 1843, John James Audubon lamented that the size of the buffalo herds was rapidly decreasing. "What a terrible destruction of life as if it were nothing," he said. "Daily we see so many that we hardly notice them more than cattle in our pastures about our homes. But this cannot last; even now there is a perceptible difference in the size of the herds, and before many years the Buffalo, like the Great Auk, will have disappeared."[137] In 1874 and 1876, Greenberg Fort introduced bills in Congress making it illegal to kill a female buffalo within the boundaries of the United States' territories. High-ranking army officers stationed in the plains during the 1870s also objected to the slaughter and forwarded letters of protest to the recently founded American Society for the Prevention of Cruelty to Animals (ASPCA), newspapers, and members of Congress. In 1874 the popular magazine *Harper's Weekly*, which had published numerous articles condemning the killing, featured "The Last Buffalo," an illustration by the artist Thomas Nast (fig. 78). Known for his merciless

THE LAST BUFFALO.

"Don't shoot, my good fellow! Here, take my 'robe,' save your ammunition, and let me go in peace."

FIG. 78

Thomas Nast
The Last Buffalo.
From *Harper's Weekly*,
June 6, 1874

caricatures of corrupt politicians and swindling tycoons, Nast took aim at the hide hunters currently at work emptying the plains. "Don't shoot, my good fellow!" the buffalo calls out as he doffs his pelt. "Here, take my 'robe,' save your ammunition, and let me go in peace."

More than ten years after Nast's "Last Buffalo" pleaded for mercy from the pages of *Harper's*, Albert Bierstadt unveiled his epic canvas, *The Last of the Buffalo* (fig. 79). The painting, the grandest bison portrayal of the century, delivered America's verdict on the matter. The artist created the vast, six-by-ten-foot canvas for display at the Paris Exposition of 1889, where the United States, England, France, Germany, Japan, and other nations exhibited works of art, natural history collections, and inventions that exemplified their highest achievements. Set in a desolate and windswept valley, *The Last of the Buffalo* describes a series of concentric circles in the chapparal. Sharp-tipped mountains ring the distance, hemming in the restless herds that churn beneath them; in the foreground, wounded bison and skulls close the visual circuit. At the center of this stark amphitheater, a buffalo and a lone Indian hunter struggle in fatal combat. A critic for the *New York World* narrated the ringside action for his readers:

With a sudden lowering of the huge head and a lunge, [the buffalo] has impaled the pony, which rears with a scream of anguish....Now toss them—ah, you were not quick enough! The yelling murderer raised that lance and shot its yard of thirsty steel downward—your heart is cloven, and horse, Indian and you drop together!...A great tragedy is the subject Mr. Bierstadt chose, and in breadth and spirit, in color and exquisite minuteness of detail, he has shown the hand of the master.[138]

Inspired by epic battles painted by Raphael, Peter Paul Rubens, Eugène Delacroix, and other European masters, *The Last of the Buffalo* hitched the story of America's bison and indigenous peoples to the glorious myths and military campaigns of Continental history. At the same time, the painting was an exposition of the evolutionary theories that were igniting the salons, pulpits, and university lecterns of Europe and the United States. A tutorial on the primacy of conflict in nature, *The Last of the Buffalo* is made up of an arena of concentric bands that allude to the chronological cycles of evolutionary change. Its mountains are the actors of geological time; the herds play out the millennia of successive generations and the fighting pair performs the span of an individual life. Consistent with nineteenth-century interpretations of Charles Darwin's researches, violence—evolution's engine—takes center stage in the painting. Bierstadt's drama pivots around the cataclysmic moment when a great evolutionary shift is finally complete. Speeding up a demise that took place over two centuries, *The Last of the Buffalo* shines a spotlight on the tragic, yet necessary, eclipse of the bison species.

Although death was the ostensible subject of *The Last of the Buffalo*, the picture enacted a ritual of birth. Viewers looking at Bierstadt's painting were claiming their right to occupy and propagate places formerly occupied by bison, indigenous peoples, and other less advanced species. According to nineteenth-century ethnologists and other scientific authorities, a defining characteristic of developed cultures, and what differentiated civilized people from savages, was the ability to create and appreciate fine art. The painting's teeming herds, which both artist and viewers knew no longer existed, pressed the victors' advantage with its bold illusion. The revived multitude showed that, while individual beings might be rooted out by Darwinian constraints, North America's great natural abundance could never be truly diminished.

Even the bones strewn across the foreground of *The Last of the Buffalo* conveyed a life-affirming message. Over the course of the nineteenth century, Americans had become increasingly anxious about the problem of "worn-out soil." Two centuries of wasteful land-use practices, including clear-cutting trees and failing to rotate crops, led to a precipitous decline in the nation's soil productivity. The large-scale, mechanized farming that became common by 1860 also accelerated the degradation of productive land, and agronomists warned of coming shortages of wheat, corn, and vegetables. When the hide hunters

FIG. 79

Albert Bierstadt
The Last of the Buffalo
about 1888, oil on canvas,
71 ⅛ × 118 ¾ in.
Corcoran Gallery of Art, Gift of
Mary (Mrs. Albert) Bierstadt

finished their work in the 1870s, it was estimated that more than two million bison had been felled. The endless sea of bison skeletons scattered over the prairie beckoned as the solution for the soil crisis and bones were dutifully collected and shipped east to be spread over depleted farmlands. The fertilizing theme of the last of the buffalo may have had a special resonance for Colonel T. J. North, the United States' "Nitrate King" who amassed a fortune selling nitrates for soil fertilizer and gunpowder. North bought Bierstadt's painting for a reported fifty thousand dollars, an astonishing amount of money that confirmed how brisk business had been.[139]

Lobbied by the nation's most accomplished artists and eminent scientists who were defining the century's extinctions as healthful natural events, Americans came to accept the idea that the buffalo had to die to allow citizens to live a better life. Even the passionate naturalist John Muir subscribed to the notion that the loss of the buffalo represented a greater good, although he deplored the brutality of the killing. "I suppose we need not go mourning the buffaloes," he wrote in 1901. "In the nature of things they had to give place to better cattle, though the change might have been made without barbarous wickedness."[140] As Bierstadt was finishing his great picture, the nation was ablaze with the news that the last of the great buffalo herds had vanished. Hide hunters still wandered the plains, guessing that the buffalo had merely migrated north into Canada and would soon return. Newspapers published unsubstantiated reports from cattle ranchers who said they had spotted bands of thirty or forty in deep brush, but the United States Army confirmed in the late 1880s that the vast herds that once had covered the prairie were no more. At last, the event that Catlin had predicted when he set out for the West in 1832 had come to pass.

While people across the country digested the fact that they would never again see herds of buffalo galloping across the frontier, a group of images from 1887 by the photographer Eadweard Muybridge showed that American ingenuity could revive, if not the bison themselves, an exciting visual facsimile. After making a series of locomotion studies of human subjects, Muybridge set up a work yard at the Philadelphia Zoological Gardens. His research menagerie included an aged bison, representative of the species that had long served as the symbol of American power. Zoologists were particularly eager to analyze Muybridge's study of the buffalo gallop, a gait that they considered

FIG. 80

Eadweard Muybridge
Buffalo: galloping
1887, collotype, $8\frac{3}{4} \times 13\frac{7}{8}$ in.
Iris and B. Gerald Cantor Center for
Visual Arts at Stanford University;
Stanford Family Collections

unique to the species and a marvel of locomotive efficiency (fig. 80). Breeders of equine stock were also interested in the gallop of the bison, which were famously able to run for ten or fifteen miles without losing speed. A traveler in the 1870s expressed the public's fascination with the bison's fastest gait. Watching a buffalo sprint over many miles, an observer from a railroad car described the buffalo's gallop as "remarkable," insisting that "one who has seen it can never forget it."[141]

Like the "formidable" stuffed buffalo that stood sentinel at the entrance to Charles Willson Peale's museum in the early years of the century, Muybridge's buffalo ushered viewers into a realm of art, illusion, and science. The picture sequence extended a lone buffalo into a herd of animals, conjuring back America's proud multitudes through the wizardry of the "automatic electro-photographic apparatus." Aboard Muybridge's image train, the bison could be "exhibited in actual

TAKING AND BEING TAKEN.

motion . . . as if the living animal itself were moving, . . . resolved into the graceful, undulating movement" of the animal bounding on forever.[142] The photographer's innovative technique was uniquely suited for both creating a vivid record of soon-to-be extinct animals and representing the progress of the United States, a country whose citizens defined themselves as "a locomotive people."

It is not known how the Philadelphia zookeepers responsible for the bison in Muybridge's photographs compelled the animal to gallop across the bank of cameras. Buffalo were notorious for their belligerence when frightened and tendency to blindly trample anything in their path. An image of 1872 depicts a mishap that befell a photographer after angering a buffalo on the Kansas prairie (fig. 81). Traveling with a scientific expedition made up of entomologists, geologists, botanists, guides, and an artist, the photographer's exploring party dispatched a few buffalo for sport then continued to hunt for fossils, flowers, and interesting views. Hearing the wild shout "Buffalo!," the group looked up to see a wounded bull bearing down on them at full gallop. After he had "ingloriously stampeded" the entomologists, the enraged animal fell upon the photographer's camera and, with a toss of his enormous shaggy head, launched it heavenward. The expedition's artist made a picture of the incident and titled it "Taking and Being Taken."

Part of the humor of "Taking and Being Taken" turns on the idea that humans have the last laugh in any contest with an animal. Still, with its flying camera representing an eye for an eye, the image is also a tale about nature taking revenge. Throughout the century, even as they energetically eliminated buffalo from the land, Americans voiced nagging concerns about the consequences of their lethal campaign. Darwin's theories offered scientific comfort that animal extinctions were inevitable and beneficial, but scripture, a far greater authority in the United States, taught a different lesson. In 1823, James Fenimore Cooper, speaking through the character of Natty Bumppo in the novel *The Pioneers*, warned that those who wantonly destroyed animals would ultimately be called to account. Bumppo watched a group of hunters indiscriminately blasting away at passenger pigeons and spoke of another observer who also was noting the tally. "The Lord won't see the waste of his creatures for nothing, and right will be done to the pigeons, as well as others, by and by."[143]

A Locomotive People

 "LIMITED EXPRESS" TRAIN IN A CURRIER AND IVES lithograph of 1884 disgorges a rabble of hurried travelers (fig. 82). The passengers' time is limited. The print shows them shoving their way through a depot door for a lunch on the fly, their "five seconds for refreshments." Since getting around a country the size of the United States took time, citizens needed to move along smartly to have a prayer of taking advantage of their democratic opportunities. Americans were a "locomotive people" and, if good manners fell by the wayside, that was of little importance to a nation speeding toward a grand destiny. Even from his leafy enclave at Walden Pond, Henry David Thoreau noticed that the railroad was accelerating the pace of life in the United States.

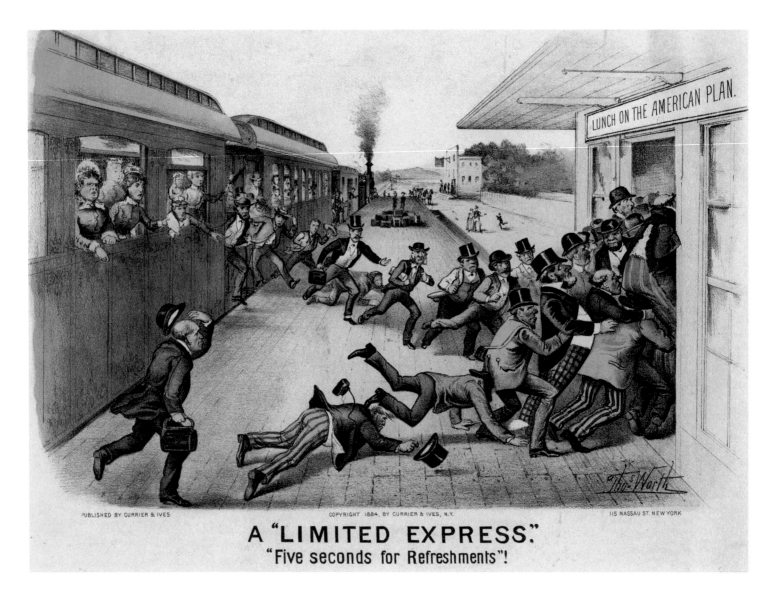

A "LIMITED EXPRESS."
"Five seconds for Refreshments"!

PUBLISHED BY CURRIER & IVES   COPYRIGHT 1884, BY CURRIER & IVES, N.Y.   115 NASSAU ST. NEW YORK

FIG. 82

Thomas Worth
Currier and Ives, publishers
*A "Limited Express.": "Five
Seconds for Refreshments"!*
1884, colored lithograph, 10 × 13 in.
Museum of the City of New York,
The Harry T. Peters Collection

"Do [people] not talk and think faster in the depot than they did in the stage-office?" he asked.[144]

The railroad emerged in the nineteenth century as the engine of American happiness, the pursuit of which was guaranteed to the citizenry by the nation's founding charter. Before the locomotive's arrival, the geographical scope of pursuing happiness was constrained by the inconveniences of the modes of transportation available. America's flibbertigibbets traveled on foot and horseback and by sailboats, barges, canoes, sleighs, and buckboards. A matter of stoic endurance, a journey on the nation's ramshackle roads required a resolute will and iron constitution. Carriages plying the routes between major cities teetered over two-foot-deep ruts in dry weather; after rain, they sank up to their axles in mud. Some towns built expensive corduroy roads made up of logs topped with sand or dirt, but the consequent lurching of wheels made carriage passengers vomit and threw them about the cabin with such force that many arrived at their destinations with broken bones and contusions. In 1800 a trip by road from Boston to Washington—one of the best-maintained routes in the country—required ten sixteen-hour days of travel. If river transportation was smoother, it also was hampered by seasonal changes in water levels. Placid waterways turned into raging cataracts in the spring, and meager flows in the summer months exposed hazardous sandbars and islands of submerged logs, defying the best efforts of river-transportation companies to maintain regular schedules.[145]

Travelers in the United States faced not only daunting distances but also the social disruptions of a country that was constantly expanding its territorial girth. In the first half of the century the United States gulped down the 828,000 square miles of the Louisiana Territory, Texas, and the vast tracts ceded by Mexico in 1848 at the end of the Mexican War; it also designated large amounts of land in the Midwest as national territory. Even as early as 1799, popular journals noted that geography books and gazetteers were unable to keep up with the "very frequent changes" in the shape and size of the United States, transformations that were "taking place daily" and a circumstance "peculiar to a new, progressive, and extensive country."[146] The rapidity with which new harbors, capes, mountain ranges, and forests were heaped onto the national plate made digestion difficult, straining the innards of a political union that already was largely an intellectual exercise. Rather than saying "the United States is," as we do today, contemporary speakers used the plural "the United States are" in recognition of a process of assimilation that was mostly incomplete. To be an American, then, was to be a reader of maps and geography books, a scout of the most recently updated information of an always-growing, plural realm.

When the English-invented steam locomotive appeared in the 1820s, Americans embraced it as the solution to the United States' troublesome spaciousness. Speed was an obsession with American railroad builders from the beginning and, pricked by the spurs of Democratic Time, they built engine lines that were fast and cheap rather than steady and durable. British railroad engineers, renowned

FIG. 83

Unidentified artist
after John B. Jervis, locomotive designer.
Design for the locomotive Experiment
which was renamed Brother Jonathan,
1831, drawing on paper. Transportation
Collections, National Museum of American
History, Smithsonian Institution,
Washington, D.C.

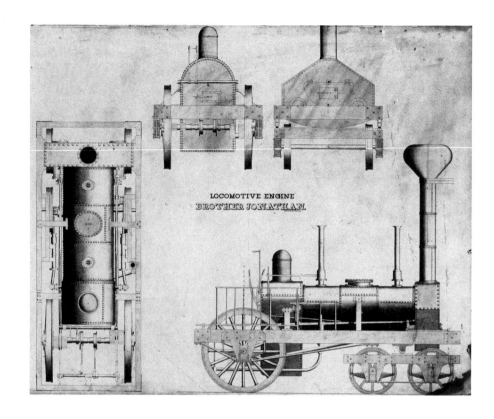

LOCOMOTIVE ENGINE
BROTHER JONATHAN.

for their devotion to durability, were appalled by the steep grades,
sharp curves, shaky wooden trestles and mountain tunnels that their
American counterparts deemed acceptable. One of the many American
engineers rushing to adapt the solidly built English locomotives to
the unstable conditions of American rails was John Bloomfield Jervis,
who tinkered with a design for a light and flexible engine (fig. 83). He
gave his new creation the name Experiment, an apt description for
an innovative model made limber by drive wheels at the rear of the
locomotive and four small swiveling wheels on a truck at the front
end. Built at the West Point foundry in 1832, Experiment (later
renamed Brother Jonathan) was a mechanical contortionist able to
cling to rickety tracks while maintaining considerable speed. Jervis's
Experiment reached the unheard-of velocity of sixty to eighty miles an
hour, making it the fastest locomotive in the world in its day. As political observers and foreign pundits were increasing their warnings about
the dangers of United States' geographical unwieldiness, the technological advances of Experiment gave evidence of a mobile nation that was
gathering momentum.

FIG. 85

Asher B. Durand
Railway Cut
engraving, $11\frac{1}{4} \times 7\frac{3}{8}$ in.
Print collection, Miriam and Ira D.
Wallach Division of Art, Prints and
Photographs, The New York Public
Library, Astor, Lenox and Tilden
Foundations

seaboard and reached from Charleston and Boston to Chattanooga, Hannibal, Milwaukee, Jacksonville, and St. Louis.

If the spark arrestors made at Baldwin Locomotive testified to the egalitarian nature of America's railroad inventions, they also gave evidence of industrial production on a scale unprecedented in the United States. The freewheeling had no place in this operational behemoth. Employees worked by a timetable and followed the rules of the companies' strict hierarchy of authority. In his *Manual for Railroad Engineers* (1878), civil engineer George L. Vose outlined the immense management edifice that governed the railroad business. At the apex of a pyramid of thousands of workers, the "Division and Branch Superintendents" watched over "Masters of Engine and Car Repairs, Car Inspectors, General Freight Agents, General Ticket Agents, General Fuel Agents, Superintendents of Telegraph, and Foremen of Bridge Repairs." This echelon, in turn, supervised "Baggage Masters, Brakemen, Conductors, Dispatchers, Engineers, Switchmen, Water Station Attendants, and Firemen." Vose recognized that the complexity of railroads presented so many variables that "it is simply impossible for science to solve them in a satisfactory manner." He advised newcomers to railroad building that "by accumulating a large amount of experience" they could "draw therefrom rules for [their] own guidance." In other words, while operators relied on standards of practical engineering and highly regulated systems of management, in the end railroading in the United States was a rough-and-ready business.[152]

Artists joined the mechanical engineers, managerial experts, and entrepreneurs in the work of inventing the railroad in the United States. If company employees directed construction and the day-to-day operation of the lines, it was the nation's painters, printmakers, and photographers who helped elicit a consensus among the citizenry about what railroads were and where they were headed. Works of art not only served to direct the location of railroad lines to scenic places where travelers, in response, soon decided they longed to go, but they also reframed the frightening jerks and blanketing soot of a train trip as part of an exciting "excursion" on a mechanical magic carpet. Asher B. Durand, who began his career as an engraver and later turned to landscape painting, produced one the earliest images of the American railroad (fig. 85). Durand, who earned a good living as a banknote engraver, in about 1840 sketched a car on a new railway line for a banknote design.

Adapting conventions of aesthetic beauty to the railroad's alterations of the land, Durand's banknote image linked bankable assets with the deeply gouged hillside of a recently made "cut." The railroad car rounds a curve on a serpentine track next to a steep rock face, just the kind of risky construction that made English railroad builders cringe. As a picture on currency that passed from hand to hand when citizens paid for postage stamps, bread, and matches, however, Durand's image linked the railroad to the essential transactions of everyday life. It also introduced the idea of a lucrative alliance between fine art and the art of railroading.

The artist Thomas Cole, a friend of Durand's and an outspoken critic of industrialization and its effects on the United States' natural beauty, warned artists of the dangers of commercial entanglements with railroads. In a poem of 1837, he declared that science and technology had "with unseemly scars deformed the land." He implicated artists in this scarring work, calling them the minions of those destroying America's landscape: "Art is captive in the gilded chains / And drags her once etherial [*sic*] form along / Soiling her wings in filth to serve for gold."[153] Cole was one of the most acclaimed painters of the period, and his criticisms were an influential indictment of the railroad just as citizens began to see it as the key to economic expansion. He entreated Americans to weigh the long-term costs of technological improvements and challenged them to consider where they wanted to go before getting aboard the railroad.

Two Artists in a Landscape, a midcentury painting by the artist Robert Havell Jr., refuted Cole's accusations with a panorama that nestled the new technology within a picturesque landscape, a common strategy by the end of the 1840s (fig. 86). Havell managed the train's unpleasant features—noise, smoke, and smell—by setting them outside the viewer's visual boundaries. Landholdings in the American countryside at that time averaged just over two hundred acres. Because most people lived or grew up on farms, that space was the distance that they were accustomed to seeing as "theirs." *Two Artists in a Landscape* honored that habit of perception by creating a sense of two hundred acres of "elbow room" between the viewer and the railway. On the left side of the image, a pair of artists sketches the scenery. One of them rummages in his color box; the other dabs at the picture on his knee as he looks out over the valley. These are not Cole's despised "captives,"

FIG. 86

Robert J. Havell Jr.
Two Artists in a Landscape
about 1840–1850, oil on canvas,
35 × 50 in. New York State Historical
Association, Cooperstown

but rather outdoor visionaries capturing with their brushes and pencils the country's exciting transformations. Certainly, painters who took the railroad as their subject were challenging a host of artistic conventions, beginning with the traditional mandate to shun parallel lines. Gilbert Stuart, an esteemed artist of the Revolutionary period, cautioned students of painting to "avoid, by all means, parallel lines whether they be parallel straight or parallel curve lines." He advised artists to examine "offensive" paintings, and if "the cause does not at once strike, bethink yourself to look out for parallel lines."[154]

Most prominent American painters deployed the two-hundred-acre strategy in their work. Setting trains at a squinting distance finessed the technical taboo against parallel lines and allowed genteel viewers to ignore the interruption of mechanical forms when looking at art that depicted natural beauty. The close-in images of trains presented in Currier and Ives's *A Limited Express* and other popular prints, on the other hand, thrust viewers into the realm of the hoi polloi, the roundhouses and train stations that were places of egalitarian jostlings, acrid odors, and imminent physical danger. An illustration appearing in *Harper's Weekly* in 1865, showing a grim reaper astride a locomotive plowing through the bodies of the maimed and dead, reminded readers that they had good reason to interpret quite literally Thoreau's comment that "we do not ride upon the railroad; it rides upon us" (fig. 87).[155]

FIG. 87

Unidentified artist
The Horrors of Travel.
From *Harper's Weekly*,
September 23, 1865

T. C. Grattan, the British consul to the United States, gave an account of one of his host nation's innumerable train accidents. On a railway journey Grattan and his fellow passengers were alarmed by a "violent jolt, accompanied by a loud crash." Because the train did not stop or slow down, the travelers surmised that all was well. At the next stop Grattan asked the engineer about the ruckus. "Well, it was in going over a chaise and a horse," explained the engineer. Shocked, the consul inquired if anyone was in the chaise, and the engineer replied that two ladies were inside.

"Were they thrown out?"
"I guess they were, and pretty well smashed, too."
"Good God! And why didn't you stop the train? Can't you send back to know what state they're in?"
"Well, mister, I reckon they're in the state of Delaware; but you'd better jump to the steamer there, or you're likely to lose your passage."[156]

Aware of their less-than-satisfactory reputation, railroad companies turned to those who created flattering images for a living. They contacted well-known painters, photographers, and print publishers and enlisted them to shift public perceptions about railway travel. Railroaders hoped to show that artists' endorsements came spontaneously, more or less, once they had experienced the wonders of train travel. In 1858 the Baltimore and Ohio (B&O) Railroad Company hatched the ingenious idea of a lavish four-day "Artists' Excursion" to scenic areas along the company's line in the Alleghenies (fig. 88). A photograph captures the delightful camaraderie of the party's June-morning departure, showing many of the nation's leading artists festooned across a shining locomotive. The party included John W. Ehninger, Thomas Hicks, John Kensett, Louis Lang, Thomas P. Rossiter, James Suydam, and a group of photographers, reporters, and writers. The hospitality of the B&O was extraordinary; the excursion train included not only a parlor car with plush couches and writing desks for taking notes, but also a dining saloon that served free-flowing champagne and hearty meals, as well as a photographic studio car outfitted with the latest equipment. Frequently stopping to allow its passengers to make sketches and photographs, the train wound its way through a gorgeous mountainous landscape blossoming into spring.

FIG. 88

Unidentified photographer
Artists' Excursion, Sir John's Run,
Berkeley Springs
1858, salt paper print, 6¾ × 6⅛ in.
Smithsonian American Art Museum,
Museum purchase from the Charles Isaacs
Collection made possible in part by the
Luisita L. and Franz H. Denghausen
Endowment

Newspapers and journals around the country reported on the Artists' Excursion. In the summer of 1859, *Harper's Weekly* featured an article in which the B&O's "Iron Horse" called upon artists to exercise their creative faculties on its behalf: "Come, ye gifted of the land," the horse said or, perhaps, whinnied.

Cease from sighing over the mouldy Past…but come, with hands of skill and hearts of fire, to glorify a Present worthy of your powers. Write, paint, sketch, and chisel that when ten and thrice ten, hundred years are gone, and our fires shall be quenched and our iron bodies heaps of rust, the noble archways that have borne us over rivers and mountain gorges shall have crumbled into ruin, the stranger (perhaps a winged tourist from some other sphere) finding a mossy stone carved with the letters B.&O.R.R. may know that they stand for "Baltimore and Ohio railroad," the grandest and most renowned work of its age! [157]

In addition to organizing journeys that motivated artists to make enticing pictures of the scenery along their lines, railroad companies also commissioned works of art directly from artists. In the mid-1850s the Delaware, Lackawanna and Western Railroad (DL&W) hired the landscapist George Inness to paint the company's line and round-house at Scranton, Pennsylvania, paying him seventy-five dollars for *The Lackawanna Valley* (fig. 89). Directed by his patrons to highlight the trains, tracks, and roundhouse, Inness uncoupled his observations from the two-hundred-acre strategies of his artistic colleagues. *The Lackawanna Valley* descended from places lofty and leafy to the hub of an industry working at full throttle. Although the scene is sprinkled with picturesque elements, including misty mountains and groves of trees on the left and right, the heart of the canvas takes up the business of removal that occupied his patrons and the nation.

The tree-stumped field in the foreground—so disturbing to modern eye—was a ubiquitous sight during the nineteenth century and one to which most citizens paid no attention. Since the arrival of the first Europeans, America's work was indeed the clearing of trees and the conversion of wilderness land into cultivated acres. Although pockets of protest erupted around the removal of large tracts of forest, most people in the United States viewed the clearing of land as a wholesome activity that dissipated woodlands' pestilential dampness and freed acreage for planting, homes, and places of worship. The cut-down trees also spoke of the railroads' craving for them; the nation's unifying system of tracks used prodigious amounts of wood for locomotive fuel and the construction of cars, tracks, and bridges. When William Penn came to the region and named the richly forested place "Pennsylvania," or "Penn's Woods," he anticipated a robust commerce in timber and placed a premium value on his settlement's standing forests. In the "Conditions and Stipulations" he laid down for newcomers, Penn specified that "in clearing the ground, care be taken to leave an acre of trees for every five acres cleared." He also appointed a state forester to safeguard the woodlands.[158] Despite the wise Quaker's careful planning, most of Pennsylvania's forested places disappeared during the nineteenth century. As Inness painted his *Lackawanna Valley*, the state's once-thriving timber industry was waning and Pennsylvania was well on its way to earning the nickname that described it by the latter part of the century, the "desert of five million acres."[159] The field of stumps

FIG. 89

George Inness
The Lackawanna Valley
about 1856, oil on canvas,
$33\frac{7}{8} \times 50\frac{3}{16}$ in.
National Gallery of Art,
Gift of Mrs. Huttleston Rogers

in *The Lackawanna Valley*, however, is a go-ahead signal, a green arrow pointing to the engine at the center of the canvas that was Scranton's locomotive heart.

If the absent trees represent one sort of removal, the locomotive in the painting that hauls a load of anthracite coal is a clue to a parallel activity taking place in the region. Pennsylvania was the focal point of the United States' coal-mining industry, and its deposits fueled much of the nation's steam-powered manufactories. Anthracite coal from mines near Scranton went to armories, textile mills, and coke-smelting operations, to steamboats plying local rivers, and to the growing number of Americans who heated their homes with it. Also a boon for train passengers, anthracite burned without making smoke or soot, and the railroad companies that used it in their hoppers made much of the fact that customers would enjoy the luxury of open windows. Although the benefits of anthracite were spread around the country, it was the residents of Pennsylvania who faced the steep payment that it exacted from the land. One contemporary observer called anthracite mining a "jealous art" and, admiring the "grand, picturesque region" of the state, rued the business that "stripped the mountains of their gaudy foliage, and leveled the venerable and sturdy forest trees to the earth."[160]

If *The Lackawanna Valley* depicted the clearings and excavations of the day, the painting's title and the name of the company that commissioned it harked back to removals of earlier days. A small number of Delaware (Lenape) people, the valley's original inhabitants, remained in the state until 1860, but most were forced out decades before by an influx of white settlers and the Indian Removal Act of 1829, which mandated that all Native Americans living east of the Mississippi move west of the Mississippi. In the Lenape language the word "lacka" referred to a point of intersection and "hanna" to a stream of water. "Lackawanna," then, may have been the Indians' term for the area where the Lackawanna and Susquehanna rivers came together, a busy place of fishing and trading.

Inness's *Lackawanna Valley* presents the junction of a new era: the roundhouse of the DL&W Railroad. A hub where the company's trains came together for repairs and switching, the roundhouse has no trees to shade it or soften its great mass. It is not surprising that Inness's patrons initially rejected the painting. The reason is unclear, but the company's directors may have been unsettled by the sight of their roundhouse clinging like a great barnacle to the land. The painting also violated the polite artistic code of the two-hundred-acre view, forcing the viewer to inspect closely the coal-hauling train, the stump field, and other trade-offs of modernization. Inness said he preferred to represent the "civilized landscape" that showed "every act of man," judging it "more worthy of reproduction than that which is savage and untamed."[161]

The Lackawanna Valley acknowledges the disorder of rapid transformation and counts on the viewer to understand that good things are coming. The figure of the red-shirted man in the painting's left foreground, watching as the scene before him unfolds, models the business of waiting that is always part of progress. The painting's utilitarian emphasis on freight en route may have disappointed the company's directors, but *The Lackawanna Valley* was, in fact, a confident endorsement of the railroad expressed in the terms of its day. It reiterated the assessment of the author Bayard Taylor who, in an article about the construction of the New York and Erie Railroad, commented on the challenge of changes and removals. "No great gift of science ever diminishes our stores of purer and more spiritual enjoyment, but rather adds to their abundance and gives them a richer zest," he declared. "Let the changes that *must* come, come; and be sure that they will bring us more than they take away."[162]

Jasper Cropsey's *Starrucca Viaduct, Pennsylvania* of 1865 pictures a landscape where the "gifts of science" and nature have reconciled (fig. 90). The Starrucca Viaduct at Lanesboro, Pennsylvania, crosses the Susquehanna River on the route of the New York and Erie Railroad. Completed in 1848, seventeen years before Cropsey painted it, the viaduct was one of the nation's grandest engineering projects and a structure that many Americans celebrated as the Eighth Wonder of the World. The bridge, constructed of hewn stone, is 1,200 feet long and 114 feet high, with seventeen solid bluestone arches fifty feet across. Trains made unscheduled stops on the way to the viaduct to allow passengers to disembark and look down on the valley. This was the panorama that Cropsey portrayed in his painting, described by a newspaper reporter as the "solid masonry of the Starrucca Viaduct, with its regular arches ... blending with the scenery the grandeur, and dignity of art, like the effect of a piece of colossal statuary in a green park."[163]

Many who saw the viaduct claimed that it conducted them back to aqueducts of ancient Rome. One of the models for the Founding

FIG. 90

Jasper Francis Cropsey
Starrucca Viaduct, Pennsylvania
1865, oil on canvas, 22 ⅜ × 36 ⅜ in.
Toledo Museum of Art, Purchased
with funds from the Florence Scott
Libbey Bequest in Memory of her
Father, Maurice A. Scott

Fathers as they mapped out the United States' republican government, the Roman Republic was also a hallowed topic in classroom primers and youth magazines. Schoolchildren imbibed the lessons of Rome's engineering feats as they learned their ABCs and practiced declaiming on the subject of its excellent roads and their unifying effect on the ancient world. If Roman engineering inspired Americans, they preferred to emulate it in quantity rather than in quality, particularly in the construction of railroad bridges. These were usually hastily built wooden affairs that routinely collapsed under the weight of locomotives; they also had a dismaying tendency to ignite and burn to the ground, thanks to the notorious "spark problem." Cropsey thus chose as his subject an extraordinary structure in the landscape of Democratic Time: a masonry railroad bridge that had provided safe passage for more than a decade and a half. Even so, viewers looking at the painting seemed to mention almost reflexively decay or collapse. A reviewer in 1866 mused, "Had a thousand years elapsed since its construction, and the work fallen into disuse and ruin, all man would be moved by its majesty and grace as they are now by the Claudian aqueduct, striding in the melancholy grandeur of decay athwart the purpled solitudes of the Roman Campagna."[164]

A picture of a rock-solid bridge that was painted in the final year of the Civil War, *Starrucca Viaduct, Pennsylvania* was also an eloquent statement about the hopes for a healing between North and South. Pictures of railroads and railroad bridges were ubiquitous features in newspaper reporting during the war. Playing a critical role in the Union victory, the North's extensive system of rails allowed it to shuttle with relative ease the troops, supplies, and armaments needed in rapidly shifting engagements. Conversely, the smaller number of Confederate lines became a critical handicap for Southern commanders. Railroad bridges, in particular, were prized military assets, and the public on both sides paid close attention to river crossings lost or gained. Every week the Northern press was filled with images of burning and exploding bridges as well as illustrated reports on Union forces' know-how in their repair and rebuilding (fig. 91). Lincoln himself complimented the army's engineers on their creativity in piecing together "bean-poles and corn stalks" to keep vital lines open.[165] The Civil War concluded with a struggle for control of critical railroad lines in the last great siege of the conflict, the battle for Petersburg.

A reminder of the final conflagrations of the war, the flaming reds and yellows of *Starrucca Viaduct, Pennsylvania* point to a change of

FIG. 91

Unidentified artist
Destruction by Rebels of the B&O
Bridge over the Potomac River,
Harpers Ferry, June 14, 1861.
From *Harper's Weekly*, July 6, 1861

season. The painting pivots around lessons reaped from pasts ancient and recent, and it couples them with a cheering vision of the "changes that *must* come" to America and the harvests of the nation's ripening industrial plenty. Cropsey exhibited *Starrucca Viaduct, Pennsylvania* as the attention of the entire country was riveted on the building of another great span: the transcontinental railroad. During the Civil War, Congress passed the Pacific Railroad Act to lay out the terms of what the *New York Times* called the "greatest undertaking of western civilization."[166] The act created the Union Pacific Railroad, which was to build west from Omaha, and the Central Pacific, which was to build eastward from California. The project entailed organizational complexities on an immense scale, formidable physical hardship for laborers, and relentless attacks by Indians armed with the latest firearms. In return for enduring the ordeal, Congress offered the railroad companies a king's ransom, including hundreds of thousands of acres of free public land and loans per mile of $16,000 for level track, $32,000 for foothill lines, and $48,000 for mountainous terrain. Since the time of the gold rush, Californians had lobbied strenuously for the transcontinental railroad, and droves of influential easterners joined the fray in the battle to make it a reality.

The respected editor Samuel Bowles made an urgent plea for the transcontinental line in his popular travel narrative, *Across the Continent* (1865):

To feel the importance of the Pacific railroad, to measure the urgency of its early completion,…you must come across the Plains and the mountains to the Pacific Coast. There you will see half a continent waiting for its vivifying influences. You will witness a boundless agriculture, fickle and hesitating for lack of the regular markets this would give. You will find mineral wealth immeasurable, locked up, wastefully worked, or gambled away, until this shall open to it abundant labor, cheap capital, wood, water, science, ready oversight, steadiness of production,—everything that shall make mining a certainty and not a chance.…You will see hearts breaking, see morals struggling slowly upward against odds, know that religion languishes; feel, see and know that all the sweetest and finest influences and elements of society and Christian civilization hunger and suffer for the lack of this quick contact with the Parent and Fountain of all our national life.[167]

After the epic transcontinental project began, the Union Pacific hired Andrew J. Russell, official photographer for the U.S. Military Railroad during the Civil War, to make a visual record of the construction. Traveling

FIG. 95

Thomas Nast after Ben Butler
The Cherubs of the Credit Mobilier.
From *Harper's Weekly*, March 22, 1873

the Central Pacific locomotive in the background that belches fire and brimstone above Stanford's head.

It is curious that Stanford chose the driving of the last spike as the subject of the painting since, by all accounts the event was a terrible embarrassment for him. Stanford missed the strike with the sledgehammer that was to mark the crowning moment of his career. A contractor who supplied beef to the construction crews and attended the ceremony recalled the "hilarious occasion" and the reaction of a drunken crowd. When Stanford blundered in his attempt to drive the last spike, he hit the rail with the hammer and a loud metallic clang pierced the air. "What a howl went up! Irish, Chinese, Mexicans, and everybody yelled with delight. 'He missed it. Yee.' The engineers blew the whistles and rang their bells."[170] Some of the assembled crew came to Stanford's rescue, and with their help he managed finally to hammer the spike in properly. A telegraph wire attached to it tapped out the news to crowds waiting breathlessly across the United States, after which three days of national festivities commenced. In Washington, D.C., a magnetic ball dropped from a pole on the Capitol building; in Chicago a parade seven miles long began, and church bells rang from San Francisco to Boston. In his portrayal of the "hilarious occasion," Hill dutifully banished the jeerers and backslappers, shuffling the facts to suit the memory his patron wished to have of the event. Yet, unsatisfied, the Governor rejected *The Last Spike* and refused to pay the artist for his five years of work on the commission. Perhaps the sting of the whistles and catcalls lingered still, and even the revisions of Hill's art could not assuage it.

An illustration from 1873 by the newspaper artist Thomas Nast suggested that the true art of railroading lay neither in feats of engineering nor in fine paintings, but rather in the creative financing that buoyed the transcontinental project (fig. 95). Nast cribbed from one of Raphael's paintings of a Blessed Virgin to portray as a pair of desultory cupids the congressmen Oak Ames and James Brooks, two of the many high-ranking officials and businessmen found to have made and accepted bribes for the transcontinental project. A year after the festivities at Promontory, a suspicious Congress launched an investigation into irregularities discovered in the Central Pacific's and Union Pacific's survey records and accounting books. Railroad surveyors for the Central Pacific, for example, had mapped out "foothills" that began in the middle of California's Central Valley. This example was just one of many creative interpretations of topography that entitled builders to the hardship premium of double or triple the amount paid for flat track.

As national headlines blazed with news—"THE ACME OF FRAUD" and "$211,299,328.17 GOBBLED"—the inquiry revealed a vast web of swindles that stretched all the way from the flats near Fresno to the U.S. Capitol. Helped by their influential friends---and probably by threats to bring others down with them---the railroad directors skirted the legal entanglements that caught up those in the lower ranks and instead spent a comfortable retirement in mansions purchased with the enormous personal fortunes they amassed in the enterprise. Meanwhile, the United States financial system, unsettled by the crushing railroad debt that the federal government and local communities would pay off for years to come, teetered into the Great Panic of 1873. As they shuttered bankrupt businesses and searched for work during the longest-lasting depression the nation had ever experienced, Americans came to associate the railroad with being railroaded.[171]

Nast's drawing affirmed that railroad builders were no angels, and it was perhaps the fall to earth of America's "moral machine" that helped give the nation its wings. Listening to their countrymen complain about railroads' despotic time schedules, fixed routes, and mass assemblies of travelers, a handful of inventors in the United States had been tinkering with flying machines since the first half of the century. One of the intrepid few was Rufus Porter, the first editor of the popular science journal *Scientific American*. During the California gold rush, Porter had designed a passenger "aerial locomotive" that could convey gold seekers to California with speed and comfort. But the entrepreneurs he approached for financial backing, preoccupied with spark arrestors and swiveling trucks, scoffed at the idea of human flight, and Rufus never found adequate funding for his airline. The printmakers Currier and Ives may have been ridiculing Porter's

FIG. 96

Nathaniel Currier
The Way They Go to California (detail), about 1849, lithograph, 10 ¾ × 17 ¼ in. Courtesy of the Oakland Museum of California, Founders Fund

FIG. 97

Augustus Koch
Bird's Eye View of Cheyenne
1870, hand-colored drawing, 18 × 24 in.
Wyoming State Archives, Department of
State Parks and Cultural Resources

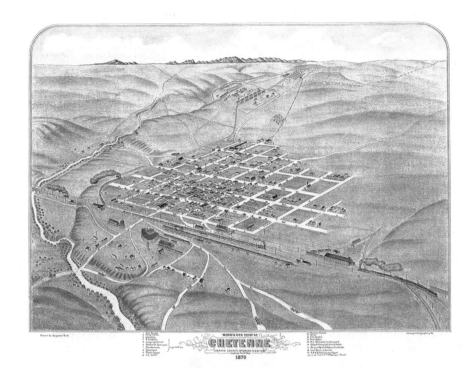

much-publicized invention when they showed an airborne gold seeker heading to the mines on a speeding rocket. "My hair!" he exclaims as he whizzes by. "How the wind blows!"[172] (fig. 96).

By the 1870s, however, America's locomotive people began to think "up." Railroad companies, eager to develop the land awarded to them as part of construction agreements, were blanketing public spaces across the country with "bird's eye" lithographs showing attractive new towns where tracts were available. Bird's-eye views were already familiar to the American public, but with the immense scale of the railroads' real estate promotions, the quality and quantity of bird's-eye views in circulation increased exponentially. Floating viewers high above the shabby storefronts and tumbleweed streets of new settlements, these aerial panoramas accustomed audiences to the visual experience of flying (fig. 97). As large numbers of people acquired the ability to imagine flying machines actually working, technologies that already had existed for decades suddenly made sense. A group of experimenters worked out the remaining refinements for a functional flying machine over the last years of the century. On a cold North Carolina morning in 1903, moving down a starting track that he and his brother Wilbur dubbed the "Grand Junction Railroad," Orville Wright took off in the first successful airplane flight in history.

[The big trees] were just sprouting when the star of Bethlehem rose and stood for a sign of its origin; they have been ripening in beauty and power through these Nineteen Centuries.

Samuel Bowles, *Across the Continent* (1865)

The Big Trees

N A SPRING EVENING IN 1850, A GOLD RUSH PROSPECTOR in California's Sierra Nevada was hunting squirrels for his dinner when he stumbled into a forest of massive trees. Looking up from cinnamon-hued trunks that were wider than steamships, he gazed in wonder at the great limbs towering three hundred feet above him. He came back the next day with his two partners and, at a loss as to what to do about their find, the band settled for carving their initials in a few of the trees. Soon the news of the state's "sylvan mastodons" spread from the mining camps to the wharves of San Francisco and the rest of the world. The newspapers of the East blazed with stories of trees as old as the pyramids and taller than the tallest buildings on earth. It was the heyday of

FIG. 98

Henry Cheever Pratt
***The Big Trees of Cal[ifornia] of the
Toulome*** [sic] ***Group from Nature***
1853, oil on canvas, 29 × 36 in.
Private collection

P. T. Barnum, however, the era when the great American pastime was sniffing out a humbug. Despite the eyewitness testimonials of numerous California authorities, most easterners dismissed the big trees as just the latest sprout in a crop of tall tales about the gold rush.[173]

While eastern skeptics passed their verdict on the mammoth trees, the survey artist Henry Cheever Pratt created *The Big Trees of Cal[ifornia] of the Toulome [sic] Group from Nature*, a visual testimonial "From Nature" and the earliest known painting of the big trees (fig. 98).

Pratt went west in 1851 as the official artist for the first U.S.–Mexico Boundary Survey and painted his picture of the Sierra forest shortly after the completion of the survey. *The Big Trees of Cal[ifornia]* retains the scientific perspective of Pratt's inventories of the Mexican borderlands; it documents the trees' vertical mass, bark texture, and foliage. To provide a key to the scale of the prodigious trees, the artist included in the painting a group of visitors on horseback dwarfed by the vast dimensions of the stand. Pratt had studied the art of landscape as an apprentice to the artist and inventor Samuel F. B. Morse, and he was well versed in inventive approaches to wilderness scenery. Artfully arranged shafts of light, moody banks of clouds, and tortured tree stumps were Pratt's stock in trade, but with plans to exhibit the canvas to doubting audiences in Boston, he fell back on workmanlike documentations. In the understated manner of a border-survey report, *The Big Trees of Cal[ifornia]* insisted that a botanical bonanza was growing in California.

Many of America's leading botanists were thrilled about the news of the big trees. As they awaited scientific corroboration, they busied themselves with the work of naming the forest giants they hoped to welcome into North America's taxonomy. They considered "Wellington" and "Washington" but ultimately settled upon the name of the renowned Cherokee chief "Sequoyah." Josiah Whitney, leader of the U.S. Geological Survey of California, explained that the *Sequoia gigantea* was "named in honor of Sequoia, or Sequoyah, a Cherokee Indian...[who] became known by his invention of an alphabet and written language for his tribe" (fig. 99). The naming of the big trees was therefore an act of remembrance for tribal people who were vanishing from the land. Whitney noted that "[Sequoyah's] remarkable alphabet is still in use, although destined to pass away with his nation, but not into oblivion, for his name attached to one of the grandest and most impressive productions of the vegetable kingdom will forever keep his memory green."[174] Native Americans were to be written out of the nation's future, in other words, but the *Sequoia gigantea* was to remain as their enduring signature.

The giant sequoias grew in seventy-five main stands in the central part of California and came into public awareness at the end of the gold rush, just as the "easy diggings" of surface deposits were depleted. For the thousands of former miners who were now looking for new enterprises, the big trees beckoned as a promise of rich farm harvests. Since colonial times, Americans had carefully appraised the size of trees on

FIG. 99

McKenney and Hall, publishers, after Charles Bird King
SE-QUO-YAH, from *History of the Indian Tribes of North America* about 1837–44, hand-colored lithograph on paper, 19 ⅝ × 13 ½ in. Smithsonian American Art Museum, Museum purchase

FIG. 100

After W. P. Blake
Mammoth Tree "Beauty of the Forest."
From *Reports of Explorations and Surveys
to Ascertain the Most Practicable and
Economical Route for a Railroad from the
Mississippi River to the Pacific Ocean.*
Vol. 5, *Report of Explorations in California*
(Washington, D.C.: Government Printing
Office, 1856). Image courtesy Yale Collection
of Western Americana, Beinecke Rare
Book and Manuscript Library

MAMMOTH TREE "BEAUTY OF THE FOREST"

properties they wanted to buy, considering tree size to be the most reliable measure of land's prospects for farming. Pictures of giant sequoias reinforced what local promoters claimed about the extravagant fertility of California's soil (fig. 100). Lined up with images of giant pears and monstrous melons in promotional journals, the big trees loomed as irrefutable evidence of a coming California cornucopia.[175]

In the context of the cooling gold rush fever, Pratt's painting of the big trees represented another kind of jackpot: a quantity of timber that beggared description and a clue of more to be discovered. Since the time of the first European voyages, North America was known for its endless tracts of "sweete forests." Wood was a precious commodity in Europe, where shipbuilders, toolmakers, and homemakers vied for an increasingly limited supply. By the time of the American Revolution, much of the wooded area of England, France, and Spain had been converted into cultivated land. The forests that remained became the domain of the monarch and the impoverished poacher—realms where the struggles of social and economic disparity played out against a backdrop of shrinking resources.

The first Europeans settling in North America encountered unbroken stands of trees that stretched beyond the horizon in every direction. Modern estimates calculate that, at the time of their arrival, a full half of the area of the contiguous United States was heavily forested, with most of the timber concentrated east of the Great Plains. Here were ship masts, roof beams, spinning wheels, church pews, and fine cabinets to be harvested over generations to come. Although much of the Atlantic seaboard forests had been cut down by 1800, nineteenth-century Americans inherited millions of acres of North American trees and the confident outlook of their forebears. For them, the discovery of the giant sequoia in California confirmed what they already knew: the timber resources of the United States were limitless. At the same time, young citizens learned at their father's knee that wooded land must be improved by its owner. A colonial poet in 1662 articulated an attitude that would prevail for two centuries when he described America's forests as a "waste and howling wilderness. / Where none inhabited / But hellish fiends, and brutish men / That Devils worshipped."[176] The forest was the haunt of the supernatural, the Indian, and the wild beast, a snare for the hapless that entered and soon found themselves "bewildered." Cutting trees was a stand against evil, an act of righteousness that wrested utility from desolation.

Seeing the exuberant forest clearing taking place all around them, William Penn, George Washington, Thomas Jefferson, John Quincy Adams, Noah Webster, and other influential Americans advised citizens to safeguard their timber resources. Tench Coxe, a protégé of Alexander Hamilton's and a leading political economist, called the

United States forests an "immense and unequalled magazine." He urged his countrymen to steward it well, cautioning that it "would be unwise . . . [for] the United States to neglect the due preservation" of woodlands. Coxe recommended a national forest policy that included government purchase of timberlands and noted that the "present redundance and cheapness of American lands enables us to effect the preservation and reproduction of our forests with less inconvenience and expense, than any other civilized nation."[177] Coxe's hopes for forest stewardship legislation, however, faced an impossibly fragmented citizenry. Kentucky trappers, South Carolina tobacco growers, Philadelphia schoolteachers, and New York financiers disagreed about everything from lawmaking and political parties to child rearing and the correct way to hitch a horse to a plow. The American electorate had reached a resounding consensus in 1829, however, when they turned their backs on the old revolutionary elites and elected the populist candidate, Andrew Jackson, whose nickname, "Old Hickory," expressed the rough-hewing mood of the country.

A decade after Jackson arrived at the White House, the English-born artist George Harvey created a group of forty watercolors portraying the "primitive forests of America." He hatched the idea of having the paintings engraved and selling them by subscription to well-to-do buyers. He titled them *Harvey's American Scenery, Representing Different Atmospheric Effects at Different Times of Day* (fig. 101). Created while Harvey resided at "Woodbank," a rustic cottage in rural New York, the images showed scenes of forest clearing with "atmospherical" effects that captured the changing light of different seasons and times of day. For the frontispiece of his folio, Harvey chose to portray the logging folk that nineteenth-century Americans lauded as "fighters of forests" felling a last few remnants. In the distance a radiant sun rises over a gleaming metropolis and a steamship in a harbor. Clouds part in the firmament above, revealing the eye of God looking on and a pair of heavenly hands clasped in a handshake. Aside from the lanky specimens spared just long enough to frame the title of the folio, trees have been cropped from the scene, yielding to homes, transportation, and commerce.[178]

While *Harvey's American Scenery* revealed the unfinished aspect of nation building, it also employed meteorological imagery to make a sunny forecast about America's future. The frontispiece resolves a cloudy sky and transient morning light into a logged landscape that also is dawning. Blending artistic sensibility, the science of optics, and a hopeful interpretation of the nation's egalitarian society, Harvey's image makes partners of axe men and art viewers in the work of a rising American civilization.[179] The resolute cheerfulness of the scene was a response to the nation's escalating arguments about forests, which had become a battleground on which differing ideas about freedom clashed. Some citizens believed that the people of the United States had the right, nay, the obligation, to create cities and towns with elm-lined streets and wooded parks where enterprising spirits could be refreshed. Others viewed such amenities as elitist affectations and claimed that all property owners were entitled to develop their land to its full extent, as they saw fit. Many people felled trees just to make that point, even when the timber served no profit. A newcomer to Michigan in the 1840s, Caroline Kirkland rallied with a group of neighbors to set aside a stand of heritage oaks for the "grand esplanade" proposed by the leaders of her small town. Kirkland wrote angrily of the "pretended blunder" that caused the trees to be the first felled by the builders. In a similar vein, the western explorer John C. Fremont told of searching for a lone tree that had for years served as a marker for travelers on the Oregon Trail. His party found it at last, cut down, unused and left to rot where it had fallen. Fremont described it "stretched on the ground . . . felled by some inconsiderate emigrant axe."[180] Perhaps the tree cutters saw old trees as a rebuke of the very newness of the United States or, in their serene immobility, as a prod to get moving. Certainly, just by enduring over generations, an ancient tree seemed to take a stand against the lickety-split schemes of Democratic Time.

In the public mind the cutting of trees went hand in hand with a host of improvements, including the propagation of new and useful plants. A portrait by Rembrandt Peale of his brother, the horticulture-minded

Rubens Peale, *Rubens Peale with a Geranium* of 1801 commented on such work and its meanings for the citizenry (fig. 102). Son of Charles Willson Peale, the famous Philadelphia naturalist and museum founder, Rubens is shown holding a potted geranium. He was captivated by plants from early childhood, and the portrait celebrates his successful cultivation of a finicky exotic, one of the first in the United States. Rubens was especially proud of his geranium, which he had coaxed from seed to flowering maturity. Even the plant-loving Thomas Jefferson had problems growing the specimen he kept at the White House.

Rubens's two pairs of glasses in the painting—the one he wears and another in his hand—concede to his poor sight as well as the method of visual classification that directed his understanding of the botanical world. The Linnaean system, used widely by botanists through much of the nineteenth century, organized plants according to their number of stamens and pistils, a first step in converting botanical resources to productive use. Rubens initially learned about the intricate morphologies of the Linnaean system from his father, who was so enamored of the method that he gave his youngest son the name "Linnaeus" in honor of the Swedish scientist who devised it. Fluent in the language of plants, Rubens knew the characteristic flower and leaf shapes of countless specimens, and his two pairs of glasses emphasize the minuteness of his Linnaean examinations. The spectacles also suggest that attention to tiny things is a correct prescription for an American citizenry always questing for the mammoth and the boundless.

Rubens is not looking at the geranium, however, and his unused pair of spectacles posits the possibility of insights unrelated to seeing. Checking the moisture of the soil in the plant's container, he gently places two fingers at its base. This feeling gesture connects Rubens to older ways of knowing plants and to practices related to medicine, healing, and magic. Such arts, often associated with female powers, relied on intuition stimulated by the engagement of all the senses. Practitioners of plant-based medicine trained themselves to see, feel, hear, and smell the subtle qualities of leaves, roots, and flowers and to formulate plant remedies according to a patient's breath, skin, pulse, and temperature. Rubens is also engaged in a kind of diagnosis, one that is rooted in an ancient human connection to the earth, a symbiotic exchange that occurs well beyond the Linnaean task of naming plants.

Joining spreading armies of apple and pear trees, grapevines, and roses, Peale's geranium participated in the nation's political, economic, and cultural declaration of independence. Over the previous century, American horticulturalists had imported thousands of new plant species as part of the struggle to create autonomy from Britain and the rest of the world. Amateur cultivators, government officials, professors of botany, and vendors in the plant trade cared most about importing edible plants and those used to make hats, carpets, rope, ardent spirits, and other products. Yet, self-conscious about the crudeness of American society, they also promoted the cultivation of exotic flowers as a genteel pursuit that disciplined the mind and elevated aesthetic sensibilities. "The pursuits of horticulture are salutary to the physical and moral nature of man," declared Zebedee Cook, vice president of the Massachusetts Horticultural Society, in 1830. "They impart vigor to the body, and expansion and elevation to the mind."[181] An accomplished painting of an expert horticulturalist, *Rubens with a Geranium* redoubles the evidence of a refining nation.

As Rubens Peale tended his Philadelphia gardens, a group of American landscape artists began to paint wild forest scenery and promote the idea that the nation's wilderness should be left in its virgin state. Their leader was Thomas Cole, a painter who worked as a maker of calico prints before emigrating from England as a youth. Cole hailed "those scenes of solitude from which nature's hand has never been lifted" (fig. 103). For Cole, it was wilderness rather than cultivated nature that uplifted the mind and spirit. He called the wild places of the United States a "sweet foretaste of heaven" and compared their "overflowing richness" to the impoverished landscapes of Europe. "[American scenery] is the most distinctive because in civilized Europe the primitive features of scenery have long since been destroyed or modified—the extensive forests that once overshadowed a great part of it have been felled. . . . The once tangled wood is now a grassy lawn." Cole extolled the "grand originality" of American forests, where "green umbrageous masses—lofty and scathed trunks—contorted branches thrust toward the sky . . . shrouded in moss of every hue and texture, form richer combinations than can be found in the trimmed and planted grove." Distraught over the widespread clearing he saw all around him, Cole pointed to the wayside that "is becoming shadeless" and imagined a ruined future. "Another generation will

FIG. 103

Thomas Cole
The Clove, Catskills
about 1826, oil on canvas, 25 ¼ × 35 ⅛ in.
New Britain Museum of American Art,
Charles F. Smith Fund

behold spots, now rife with beauty, desecrated by what is called improvement; which, as yet, generally destroys Nature's beauty without substituting that of art."[182]

The task of smoothing away the bumps on the route to an improved future fell to Cole's heirs, the artists of the Hudson River school who specialized in portraying wild scenery that was accessible to ramblers and excursionists by foot, carriage, and train. Their paintings of granite crags, frothy waterfalls, and woodlands clothed in autumn colors labored to resolve the rising ambivalence about the toll of development, inventing a United States where nature and human purpose coexisted in harmonious balance. *View toward the Hudson Valley*, created in 1851 by Asher B. Durand, embodies the nation's flourishing art of landscape

FIG. 104

Asher Brown Durand
View toward the Hudson Valley
1851, oil on canvas, 33⅛ × 48⅛ in.
Wadsworth Atheneum Museum of Art,
The Ella Gallup Sumner and Mary Catlin
Sumner Collection Fund

in both its painterly and nation-building aspects (fig. 104). On the left of the canvas, two men converse amiably as they emerge from the woods. They are strolling in the direction of a tidy village that has supplanted the forest that once grew there. In a meadow, between the woods and a white church spire in the distance, grazing cows signal that a process of digestion is under way.

The two walkers are mirrored by a pair of fine beeches in front of them. One of the men gestures toward the trees and the clearing ahead, as if imparting important information to his companion. Is he commenting on the valuable instruction that trees offered regarding democratic and moral principles? That was a frequent topic in nineteenth-century American botany texts. A typical example, *What May Be Learned from a Tree?* (1860), expounded on the question of its title for more than twelve chapters. The author, botanist Harland Coultas, explained that trees gave the people of the United States "clear and comprehensive views of the organization and laws which govern the civilized world!" They also served as teachers of the "rules of conduct which lie at the foundation of all success in business, all progress in the pathway to preeminence." Envisioning trees as models of national unity, Coultas called his readers' attention to the "few leaves [which] by their united labors form a shoot; and this, by repetition of itself, has produced a great tree."[183] *View toward the Hudson Valley*, with its bisecting "pathway to preeminence," is also a lesson from the forest. With the two men who pause at a spot between the woods and the town, the artist builds his painting around the place of balance between the energies of nature and those of settled industry.

Durand made trees his artistic specialty, and his beeches, chestnuts, and elms graced the morning rooms of wealthy industrialists and hung in fashionable New York galleries. A sophisticated urbanite who never lived more than twenty miles from the business center of Manhattan, he was a sought-after art instructor who counseled his students to go forth into the woods. A portrait of Durand painted in 1857 by his friend Daniel Huntington shows the artist doing just that, depicted here at the groves of Franconia Notch, one of the nation's celebrated vistas (fig. 105). Audiences praised the botanical accuracy of Durand's paintings, and both botanists and amateur plant enthusiasts used them to memorize the orbicular, ovate, ciliate, lobed, toothed, and hairy leaf shapes described in botany texts. Durand maintained that tree observations were the foundation of artistic training. He directed the would-be painter to "observe particularly wherein [the tree] differs from those of other species: in the first place, the termination of its foliage, best seen when relieved against the sky, whether pointed or rounded, drooping or springing upward, and so forth; next mark the character of its trunk and branches, the manner in which the latter shoot off from the parent stem, their direction, curves, and angles."[184]

Paintings by Durand and other Hudson River artists vouched that the United States had more trees than its

FIG. 105

Daniel Huntington
Asher Brown Durand
1857, oil on canvas, 56 ⅛ × 44 in.
The Century Association, New York

FIRST WEEK
Of the Great Exhibition of the
MAMMOTH TREE
AT
No. 596 Broadway
Immediately adjoining Niblo's.

THE GIGANTIC REPRESENTATIVE
OF THE
California Forests
May be seen as above during the
DAY AND EVENING
Between the hours of 9 A. M. and 10 P. M.

ADMISSION 25 CENTS.   CHILDREN - HALF-PRICE

THIS IMMENSE TREE
IS OVER
3,000 YEARS OLD

Its prodigious dimensions are without parallel in the Vegetable
Kingdom. While standing in its native Region, it measured

325 FEET IN HEIGHT

Add one-third to the height of Trinity Church Steeple, or to Bunker
Hill Monument, and the dimensions of the main trunk of this tree
would be about the same. Fifty feet of the bark, from the lower
part of the trunk, has been put in the natural form, and is now on
Exhibition at 596 Broadway, where it forms a spacious Carpeted
Room,

92 FEET IN CIRCUMFERENCE

Inside of which is a Pianoforte and Seats for 40 persons. Thirty
two Couples have waltzed within its enclosure, without difficulty.

This Vegetable Monster

When felled, contained from actual measurement, 300 Cords of
Wood. It occupied the labor of ten men for twenty-six days in
felling it, which was effected by boring through the immense mass
with pump augers.

A piece of wood will be shown, which was cut from the tree
across the whole diameter. In this piece will be seen the concentric rings denoting its great age, and also several charred places
near its centre, showing where fires had been kindled against its
trunk many centuries ago, by the aboriginals.

N. B.---This tree has never before been exhibited in this City,
nor is there any other of even one-third its size in this Country.

New York City
1854

people could ever fully utilize. It was no wonder, then, that audiences in 1855 gave a lukewarm reception to the "Big Tree on Broadway" that had sailed, round the Horn, to New York. The specimen was not an entire tree; rather, it was the bark of a giant sequoia that had been stripped to a height of 116 feet and reassembled in sections in a Broadway exhibition hall. The show was the brainchild of a pair of business-minded Californians who recognized a lucrative opportunity when they came upon the mammoth redwoods of the Calaveras Grove. After purchasing the "Mother of the Forest" from its owner for one thousand dollars, the duo worked for three months to debark the "vegetable monster," then lashed it in the hold of a steamship and sent it to the East. Broadsides and advertisements in the *New York Herald* provided the necessary statistics (fig. 106). "This gigantic monarch of the forest measured, while standing, three hundred twenty feet in height, and ninety five feet in circumference," the *Herald* proclaimed. "It was first exhibited at San Francisco, where the interior formed a spacious carpeted room, containing a piano, with seats for forty persons. . . . On one occasion one hundred and forty children were admitted, without inconvenience; and, at another time, thirty two couples waltzed within its colossal enclosure."[185]

The advertising fanfare that accompanied the giant sequoia's debut in Manhattan did not convince many New Yorkers. One of the promoters complained that "hundreds of people would come in and buy a ticket, look the bark and tree and wood over and remark that it was impossible that such a thing ever grew and that they were being humbugged."[186] Horace Greeley, editor of the *New York Herald*, tried to stir up some excitement by comparing the big trees with the wonders of the ancient world. "That they were of very substantial size when David danced before the ark, when Solomon laid the foundations of the Temple, when Theseus ruled in Athens, when Aeneas fled from the burning wreck of vanquished Troy . . . I have no doubt."[187] Nevertheless, eastern audiences were unconvinced and soon forgot about the big trees. There was little question of actually visiting the giant sequoia forest groves, tucked away in the remote recesses of the Sierra Nevada. Before the completion of the transcontinental railroad in 1869, traveling to San Francisco involved mammoth unpleasantness: a six-week sail to San Francisco via the malarial swamps of Panama, a stomach-churning passage around Tierra del Fuego, or a three-month overland excursion through hostile Indian territory, deserts, and high mountain passes. Earning enough revenue from "Mother" to return to California, the tree's disappointed owners found that its giant stump was a draw for a growing stream of tourists. They constructed a saloon and two bowling alleys on top of it and carved off sections to make some five thousand souvenir walking canes for visitors—a paltry recompense given the size of the amputation.

In 1862 the giant sequoias were again on view in New York, but this time the American public was riveted by what they considered to be irrefutable scientific proof of the trees' existence. The photographer Carleton Watkins, who had traveled to Yosemite in 1861, displayed a group of his California photographs at the elegant Goupil's Gallery. The exhibition caused a sensation, and the cream of society thronged to see pictures of the sublime places at the edge of the continent: Inspiration Point, Bridal Veil Falls, Half Dome, and the fabled mammoths of the Mariposa Grove (figs. 107 and 108). Here at last, in a scientific medium produced by a machine that could not exaggerate or lie, was evidence that the great forest realms of California truly existed. Ralph Waldo Emerson joined the enthusiastic crowds who saw the photographs and was one of many illustrious Americans who wrote about them.

Formerly skeptical, Emerson declared that Watkins's photographs "make the tree possible."[188]

Part of what made the tree possible for Emerson and other viewers was their knowledge of the laborious technical process that Watkins's photographs required. Audiences familiar with photography's complicated chemical protocols were informed that Watkins traveled to the Sierra Nevada hauling more than one hundred glass plates, each weighing nearly four pounds, in addition to processing trays, chemicals, a dark tent, and a stereoscopic and a mammoth plate camera as well as tripods to hold them.

Confronted by the technical tours de force of Watkins's pictures, even scientists scrambled to make sense of them. The top echelon of the American scientific community, including William Brewer and Josiah Whitney of the California Geological Survey, botanist Asa Gray, and Harvard zoologist Louis Agassiz scrutinized the photographs' wealth of details. Agassiz praised the photographs as extraordinary scientific documents, calling them the "best illustrations I know of the physical character of any country."[189]

In true American fashion, it was the science of statistics rather than botanical science that framed the public's understanding of the giant sequoia. An emerging field of studies that Thomas Jefferson and other Founding Fathers considered to be of vital importance for democratic governance, statistics promised equitable representation and lawmaking through tabulations and comparisons regarding the people of the United States, their churches, schools, wages, and occupations, as well as accountings of the country's natural resources. Nineteenth-century Americans were a statistical people committed to gathering and interpreting data about what President Martin Van Buren called their "longevity, social happiness, [and] domestic happiness."[190] Accordingly, the nation's first step toward understanding the big trees was to count everything about them. State reports and tourist guidebooks converted each feature into a number: heights, circumferences, the number of rings in a stump, the number of feet to a first branch, the depth of bark, the weight of cones, and the span of limbs. Comparisons between noteworthy specimens made up a significant portion of such reports, creating the impression of titanic rivalries. The 435-foot "Father of the Forest" prevailed over the 320-foot "Hercules," and the broken-topped "Old Maid" was clearly out of the running at a puny 261 feet.

Watkins's California photographs appeared in the East as the Civil War was beginning, and the trove of natural history statistics that accompanied them offered a welcome respite from the growing tally of the dead, wounded, and missing. The photographs generated such excitement about the wonderful places on the Pacific that their message ultimately reached the U.S. Congress and the White House. In 1864, Abraham Lincoln turned away from the work of war to sign legislation that set aside the Yosemite Valley and the Mariposa Grove as a state preserve. With this relatively minor act of preservation—just ten square miles—the federal government gave the power of law to the idea that intact nature had value. It is not known if congressmen, caught up in the hard fight against the Confederacy, took a special interest in the giant sequoia because some of the trees bore the names of esteemed wartime personages, including Abraham Lincoln and General Grant. Undoubtedly, the Battle of the Wilderness in 1864, one of the bloodiest engagements of the Civil War, was on every American's mind at the moment of the legislative decision. Clashing for three days in a dense Virginia forest of pine and scrub oak known as "The Wilderness," 18,440 Union and 11,400 Confederate soldiers lost their lives. As America's men fell, national leaders resolved that a distant grove of trees in California would remain standing.[191]

Congress preserved Yosemite in recognition of its scenic beauty, but the Yosemite Act also spoke for those who believed that healthy forests represented incalculable practical benefits. The outspoken champion of these ideas was George Perkins Marsh, a polymath Vermont congressman, lawyer, lumber dealer, diplomat, farmer, and businessman. In 1864, Marsh published *Man and Nature; or, Physical Geography as Modified by Human Action*, a book that helped set in motion an accelerating American awareness of the economic and social value of intact ecosystems. Like a photosynthesizing plant, Marsh took in the disparate intellectual energies of historians, hydrologists, philosophers, botanists, agronomists, poets, and geologists and converted them into an overarching canopy of ideas about nature. He showed that forests, climate, rivers, animals, and human beings did not exist separately, but rather as constituents of an interdependent whole—a call for unity that had special meaning for Americans in the darkest hour of the Civil War. Marsh hoped to raise an alarm in the United States by describing the fate of the ancient civilizations of the Mediterranean, which disappeared after empire builders cut down the forests that maintained soil fertility and a climate hospitable for agriculture. Describing the richness and beauty of once-wooded lands in Asia Minor, Greece, and Alpine Europe, he wrote of regions that had been transformed into places of "desolation almost as complete as that of the moon."[192]

Marsh beseeched his readers to "profit from the wisdom of our older brethren" and called on them to preserve "American soil . . . as far as possible, in its primitive condition." Electrified by Marsh's recommendations, landscape architect Frederick Law Olmstead, justice Oliver Wendell Holmes, poet and editor William Cullen Bryant, and other distinguished Americans began to campaign vigorously for the preservation and restoration of the nation's woodlands. Bryant, poet of the wilderness sublime, echoed Marsh's teachings about nature's utility when he wrote, "Thus it is that forests protect a country against drought, and keep its streams constantly flowing and its wells constantly full."[193] *Man and Nature*, the Yosemite Act, and the union of the poetical and the practical in Bryant's words marked a watershed moment in the United States, a juncture when the nation's builders paused to consider what they had wrought.

In 1864, the year that Marsh published *Man and Nature*, Winslow Homer presented a tree as a protagonist in a haunting Civil War painting. *The Initials* tapped into a reservoir of emotions related to the war that were uppermost in the national psyche that year (fig. 109). A large pine dominating the center of the canvas is paired with an elegantly dressed young woman who presses her fingers to unseen initials carved in its bark, not unlike the young Rubens Peale checking on his geranium. The crossed cavalry swords incised on the front of the tree serve as a grim reminder of men cut down in other forests, and the woman makes a connection to them with her gentle touch. Like the marks made by nineteenth-century surveyors on "witness trees" to confirm the location of a boundary, the initials suggest a place of crossing over. The woman's clothing also speaks of a threshold. She wears a bright blue gown with black trim on the bodice and sleeves, and a straw hat with a black band, a suggestion of the "half-mourning" dress worn by grieving women after a year of the black crapes, shawls, and veils of full mourning. Poised between the worlds of the living and the dead, the rememberer in the forest thinks of someone who has departed.[194]

Created by the artist during the darkest hour of the war, *The Initials* left the horrors of battle out of the picture. The painting seemed to evoke instead the "ministering angels" of the long campaign, the female nurses who bandaged, bathed, and fed wounded soldiers. In a conflict where one out of six wounded died owing to inadequate medical staff and unhygienic conditions, the aid of these middle-class ladies was of crucial importance (fig. 110). Their presence alone helped boost the morale of troops enduring the impossible. Nurses rallied around the inspirational Florence Nightingale and Clara Barton, who were hailed as national heroines for organizing hospitals and volunteer services. Nightingale, from England, was best known for her services during

FIG. 110

James Gardner
Nurses and Officers of the United States Sanitary Commission at Fredericksburg, Virginia, during the Wilderness Campaign
1864, collodion print.
Library of Congress

the Crimean War, but the naming of a tree at the Mariposa Grove in her honor in 1862 was just one of many tributes bestowed on female nurses for their assistance to America's fighting men. Barton explained of her war work, "I may be compelled to face danger, and never fear it, but while our soldiers stand and fight, I can stand and feed and nurse them."[195]

Although *The Initials* deals with a Civil War subject, it also draws upon other conflicts in American society and healing of a broader nature. Homer's protagonist, reading what a tree says, stands for the generations of American female plant specialists. These were farm women who gathered leaves in the woods to make potions for their families' ailments, and middle-class "botanizers" who tramped in long skirts over hill and dale to claim botany as a viable realm of female endeavor. The woman in the painting also represents female authors, clubwomen, and educators who were among the most outspoken opponents of indiscriminate logging and whose protests predated by decades the warnings of George Perkins Marsh. The author Lydia Sigourney articulated the feelings of many of her countrywomen when she declared in a poem of 1845, "Fallen Forests," "Man's warfare on trees is terrible." The poem was a eulogy for "the loftiest trunks, that age on age / Were nurtured to nobility, and bore / Their summer coronets so gloriously." The poem describes them, finally, as trees that "Fall with a thunder-sound, to rise no more."[196]

Woodlands near settled areas often provided a small independent income for women in a society that offered few decent-paying jobs for females. Even when busy with their appointed chores as the nation's homemakers, women went to the woods to collect chestnuts, berries, mushrooms, and other delicacies to sell on their doorsteps. They applied the earnings to their various aspirations: education for their children, home improvements, a measure of security in old age. Female writers and educators also defined trees as an integral part of homemaking, viewing them as essential to the pleasant and healthful communities where they wanted to live. In *Rural Hours* (1850), Susan Fenimore Cooper recognized that "in planting a young wood, in preserving a fine grove, a noble tree, we look beyond ourselves to the band of household friends, to our neighbors." In 1864, when American lawmakers marked the Mariposa Grove as a place of amnesty, they had attuned at last to a chorus of female voices.[197]

A decade later, in 1874, the Civil War was long over and a new transcontinental railroad spanned the nation from coast to coast. A trek to the land of the big trees was no longer reserved for the halest of men but instead was a journey of "rapidity and ease" and a "triumph of well-directed science and mechanics." Artists flocked to the giant sequoia groves, eager to capture with their paintbrushes the wonderful subjects of Carleton Watkins's photographs. The most celebrated painter to make the trip was Albert Bierstadt, famous across the nation for his portrayals of big things: big buffalo herds, big mountains, big western skies. Bierstadt had visited California in the 1860s, traveling with the art critic Fitz Hugh Ludlow, who warned that the gargantuan dimensions of the giant sequoia made them exceedingly difficult to portray. He declared that artists in California "neither made nor expected to make anything like a realizing picture of the groves" and he identified the painterly problem. "The marvelous of size does not go into gilt frames," he explained. "You paint a Big Tree, and it only looks like a common tree in a cramped coffin." According to art critics in the

East, most paintings of giant sequoia were dismal failures, and they pronounced the trees unpaintable.[198]

Bierstadt began his Sierra foray in 1874 by making a group of detailed sketches, then retreated to his studio in the East to attempt what all said could not be done. He emerged with *Giant Redwood Trees of California*, a painting that abandoned the conventional and much-criticized tall, narrow format (fig. 111). Humbly, as if admitting that his artistic talents were no match for the sublime verticality of the Mariposa Grove, the artist used a broad rectangle measuring 52 ½ by 43 inches and depicted only the lower part of the trees. He allowed his viewers to imagine what is left out—a canopy where swaying boughs weave strands of sunlight, shifting clouds that threaten rain, a rising distant peak, or other signature features they knew well from Bierstadt's acclaimed paintings of the western landscape. It is curious and telling that *Giant Redwood Trees of California* cut off the trees at the height of the stripped sequoia bark that was exhibited on Broadway in the 1850s, as if calling attention to its function as an artistic illusion, or a humbug.

In 1869 the influential art critic James Jackson Jarves described the "vulgar ideas of 'big things' " in Bierstadt's paintings' as a kind of swindle, insisting that "the return to the spectator who thinks, or has the spiritual faculty, is not worth the cost." He insisted, "In reality, [Bierstadt's works] are bold and effective speculations in art on the principles of trade."[199] Perhaps it was the same enterprising spirit that Jarves detected in Bierstadt's paintings that led the artist later, in 1896, to design and patent an "expanding railway car" to convey passengers in comfort on transcontinental railroad journeys (fig. 112). Sharing the approximate dimensions of the windows in the artist's patented railway car, *Giant Redwood Trees of California* offers a scenic view of the realms of ancient stories: gardens of Eden, Druid precincts, and places of merry men. Here, the art of landscape becomes a smooth transport, making a stop at a depot of fantasy and "speculation."

The painting includes a band of Native Americans that points to a very different art of landscape, however (see detail, p. ii, frontispiece). Approaching a pair of men, a female Indian carries an acorn-gathering basket on her head. The band embodies the work of tilling, planting, pruning, and harvesting through which tribal people created an environment that had nurtured hundreds of generations in California. The

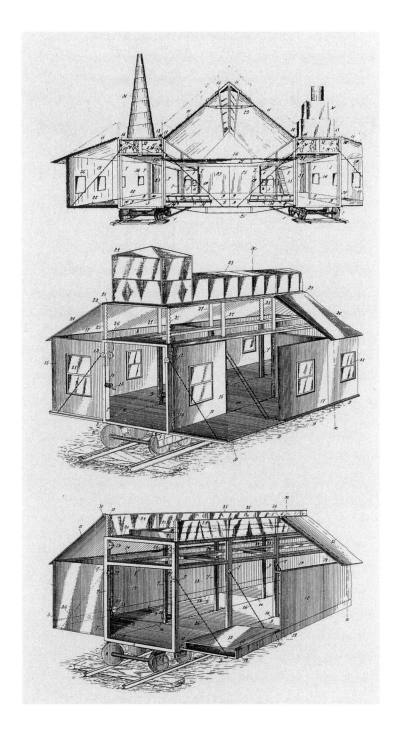

FIG. 112

Albert Bierstadt
Patent drawing for the Expanding
Railway Car, 1896. National Archives
and Records Administration

looming giants around them are totems of the Native Americans' immense repository of botanical knowledge, built up over millennia. When nineteenth-century scientists counting the giant sequoias' thousands of concentric rings began to speak of the pyramids and the Parthenon, they might also have recognized the enduring landscape art of the tribes that thrived in the land of the big trees. The whole of California was their great canvas, made up of carefully tended habitats that supported some of the highest-density populations in North America and some of its most diverse societies. Experimenting in their natural laboratory, the land's original people devised husbandry systems, pharmacopia, and building practices sustained by their wide-ranging knowledge of nature.[200]

In fact, the scenery portrayed in *Giant Redwood Trees of California* was not a wilderness but a domesticated landscape. Long before the arrival of the first Europeans, Indian horticulture had created the open Yosemite Valley, and its tribes had carefully cleared and burned the underbrush in the Mariposa Grove. The interventions by native people in California, which included control of river flows, plant cultivation, and strategic burnings, took place on such an immense scale that many of the area's ecosystems developed as a result of them. Indeed, the health of coastal prairies, black oak savannas, and dry montane meadows depended on human tending. These gardenlike botanical communities, which early visitors found so remarkable and beautiful, were regenerated over time by residents who had learned how to ensure an abundance of plant and animal life through their strategic provisioning. A landscape-shaping people, Native Americans left their initials throughout the California landscape.[201]

Bierstadt's vision of ancient trees and people begged a question of the newcomers who were its audience: what were they cultivating, for whom, and for how long? *Giant Redwood Trees of California* and the artist's expanding railway car testified to a growing web of businesses and industrial schedules that now stretched from the Atlantic seaboard to the shores of the Pacific. Although the big trees of the Mariposa Grove were protected, energetic logging was under way at similar stands in other parts of the state. The purpose behind the cutting was unclear, since giant sequoia wood is brittle and ill suited for construction. In any case, settlement in the region was low, so the wood was usually shipped to agricultural areas to be used for grapevine stakes, trays for drying raisins, and other trifles. In 1847, Emerson commented on the results of the United States' accelerated harvesting methods. "Alas for America as I must so often say, the ungirt, the diffuse, the profuse, procumbent, one wide ground of juniper out of which no cedar, no oak will rear up a mast to the clouds!" he mused. "It all runs to leaves, to suckers, to tendrils, to miscellany. . . . America is formless."[202] From the vantage point of the twentieth century, a Southern Sierra Miwok elder looking at the Mariposa Grove and other lands that had been set apart from human tending seemed to confirm Emerson's predictions. "It's turned back to wilderness," he said.[203]

Epilogue

In 1859 the Ohio artist Robert Duncanson created *Landscape with Rainbow*, a painting of a sunlit meadow in a broad valley (figs. 113 and 114). Framed by stately trees that rim the landscape, cattle meander by the side of a river where a barefoot girl and boy have stopped to talk. The children gaze at a rainbow appearing in the sky, a hint that a storm has recently passed. Touching down on the far right of the canvas, the rainbow's luminous vapors mingle with the chimney smoke of a humble cabin nestled in the woods. The glowing arc blesses the rustic home and envelops the newly settled valley in a halo of enchantment. Duncanson's landscape is a place of beginnings, with wildflowers blooming, a young generation coming up, and, in the cabin, a family starting afresh. The rainbow also extends its luminous precinct to places beyond our view, encompassing in its span other dwellings across the countryside. A picture of a point of departure, *Landscape with Rainbow* captures a moment when newcomers were thinking of the future and deciding what the land would become. The rainbow embodies the promise of the valley, as well as the fleeting nature of such a scene in the enterprising United States. Throughout the nineteenth century, landscape painters had visualized the kinds of places where patrons and viewers said they wanted to live. Duncanson portrayed in his painting a realm poised between the wilderness and the crowded city, the middle landscape that America's founders, poets, and average citizens defined as the foundation for happiness, prosperity, and freedom.

As an acclaimed African American artist, Duncanson was a wonder in his time. During much of the nineteenth century the doors of the art salon, the company boardroom, and the university were closed to black Americans. At midcentury most were slaves on Southern plantations, but even free blacks in the North were confined to the most menial jobs as farmhands, porters, and chimney

sweeps.[204] Duncanson envisioned something different and, outfitted with his brush and tubes of paint, he reconnoitered the boundaries of what was considered possible in the United States. Among his many paintings, *Landscape with Rainbow* was an especially daring foray since it represented a natural phenomenon that nineteenth-century artists considered one of the most difficult subjects to pull off convincingly. Many a wooden gangplank transited the canvases of the period, and art critics delighted in skewering the makers of inept rainbows. Furthermore, as Duncanson worked out his composition, national audiences were swooning over the breathtaking rainbow in Frederic Edwin Church's *Niagara* of 1857 (fig. 36, p. 64–65), which reviewers had pronounced the last word on the subject.[205]

With *Landscape with Rainbow*, Duncanson placed his painterly interpretation alongside the highest accomplishment of an American master and, despite the quagmire of racism that bounded it, contemporary critics hailed the painting as a resounding success. One reviewer deemed it "most expertly managed," while another mused that he wished that its landscape were a real place where he could go. "We felt as if we could make almost any sacrifice to lay down by such a river and dream of that truth and beauty which the mind never evokes but in close proximity to nature's unsullied beauties." In 1861, when the *Daily Cincinnati Gazette* praised Duncanson as "the best landscape painter in the West," it also recognized the artist's status as an inventor, that most esteemed of American professions.[206] Duncanson's creation had nothing to do with cams, levers, or patents, of course. In claiming his right to participate in the nation's most elevated endeavors, the intrepid artist had invented himself.[207]

The signature of his artistic attainments, Duncanson's well-realized rainbow also placed his painting solidly within the conversations of science at a time when Americans believed there was no question that science could not answer. Nineteenth-century authorities in optics and physics were fascinated by the heavenly bands of color and attempted to unravel their mysteries. Although scientific theories abounded about the mirror, lens, and prism effects of water droplets in the air, none provided a complete explanation for the phenomenon. A work of art that melded factual and inspired ways of seeing, *Landscape with Rainbow* suggested that science alone was not equipped to explain wonderful things. The painting anticipated theories posited by modern researchers who, still perplexed by the rainbow, propose that one part of our vision of the arching spectrum takes place in the sky, and the other, within our minds. The splendor of the rainbow, in other words, is both a matter of physics and an act of imagination.[208]

The idea that a true understanding of nature comes to light where creative vision and a rooted sense of reality converge was what Charles Willson Peale wished to impart to visitors at his Philadelphia museum. He stocked his galleries with scientific evidence of the sheer abundance of America's offerings, but the order of his collections was predicated on much more than a bonanza of natural resources awaiting consignment. With reverence, the artist laid out a project of stewardship and invited citizens to devise a nation that would endure. A laboratory of collaboration and goodwill, Peale's hall of wonders posed the question that first sparked the Great Experiment: can you imagine?

FIG. 113 (opposite)
William Notman
Robert S. Duncanson
1864, albumen print, $3\frac{1}{2} \times 2\frac{1}{4}$ in.
Notman Photographic Archives,
McCord Museum of Canadian History

FIG. 114

Robert S. Duncanson
Landscape with Rainbow
1859, oil on canvas, $30 \times 52\frac{1}{4}$ in.
Smithsonian American Art Museum,
Gift of Leonard and Paula Granoff

❧ Notes

PREFACE

1. In 1790, a year after his inauguration, Washington again described the United States as an experiment when he wrote that the "establishment of our new Government seemed to be the last great experiment for promoting human happiness." Remembering Washington after his death, Thomas Jefferson recalled that the United States' great general and first president had "often declared to me that he considered our new constitution as an experiment on the practicability of republican government, and with what dose of liberty man could be trusted for his own good; that he was determined the experiment should have a fair trial, and would lose the last drop of his blood to support it." On Washington's ideas about what continued to be popularly known as the Great Experiment well into the nineteenth century, see John Rhodehamel, *The Great Experiment: George Washington and the American Republic*, exh. cat. (New Haven, CT: Yale University Press; San Marino: Huntington Library, 1998).

2. Many studies cover the subject of science and wonder, including James Delbourgo, *A Most Amazing Scene of Wonders: Electricity and Enlightenment in Early America* (Cambridge, MA: Harvard University Press, 2006); Stephen Greenblatt, "Resonance and Wonder," in Ivan Karp and Steven D. Lavine, eds., *Exhibiting Cultures: The Poetics and Politics of Museum Display* (Washington, DC: Smithsonian Institution Press, 1991), 42–56; Richard Holmes, *The Age of Wonder: How the Romantic Generation Discovered the Beauty and Terror of Science* (London: Harper Press, 2008); and Fred Nadis, *Wonder Shows: Performing Science, Magic, and Religion in America* (New Brunswick, NJ: Rutgers University Press, 2005).

3. The publications that have inspired many of the ideas in this book include Robert Axelrod, *The Evolution of Cooperation* (New York: Basic Books, 1984); Janine M. Benyus, *Biomimicry: Innovation Inspired by Nature* (New York: Morrow, 1997); Christopher Boehm, *Hierarchy in the Forest: The Evolution of Egalitarian Behavior* (Cambridge, MA: Harvard University Press, 1999); Mihaly Csikszentmihalyi, *Creativity: Flow and the Psychology of Discovery and Invention* (New York: HarperCollins, 1996); Gretchen C. Daily and Katherine Ellison, *The New Economy of Nature: The Quest to Make Conservation Profitable* (Washington, DC: Island Press/Shearwater Books, 2002); David Edwards, *ArtScience: Creativity in the Post-Google Generation* (Cambridge, MA: Harvard University Press, 2008); Howard Gardner, *Multiple Intelligences: The Theory in Practice* (New York: Basic Books, 1993); Paul Hawken, Armory Lovins, and L. Hunter Lovins, *Natural Capitalism: Creating the Next Industrial Revolution* (New York: Little, Brown, 1999); Evelyn Fox Keller, *Making Sense of Life: Explaining Biological Development with Models, Metaphors, and Machines* (Cambridge, MA: Harvard University Press, 2002); Bill McKibben, *The End of Nature* (New York: Random House, 1989); Pavan Sukhdev, *The Economics of Ecosystems and Biodiversity: An Interim Report*. (Bonn, Ger.: United Nations Environment Programme, 2008); and Donald Worster, *The Wealth of Nature: Environmental History and the Ecological Imagination* (New York: Oxford University Press, 1993).

THE HALL OF WONDERS

4. John Davidson Godman, *American Natural History*. Vol. 1, pt. 1, Mastology, 1826–28 (rpt., New York: Arno Press, 1974), 208. For details of Peale's "mammoth feast," see "American Miracle," *Poulson's American Daily Advertiser*, February 18, 1802, 2. On the cultural history of the American mammoth in the early United States, see Paul Semonin, *American Monster: How the Nation's First Prehistoric Creature Became a Symbol of National Identity* (New York: New York University Press, 2000).

5. The observer was Peale's eldest son, Rembrandt Peale. Quoted in John C. Greene, *American Science in the Age of Jefferson* (Ames: Iowa State University Press, 1984), 286.

6. Numerous insightful scholars have described Peale's discovery of the mammoth. Some of the sources that have guided my comments on this subject include Abraham Davidson, "Catastrophism and Peale's 'Mammoth,'" *American Quarterly* 21, no. 3 (1969): 620–29; Lillian B. Miller, "Charles Willson Peale as History Painter: *The Exhumation of the Mastodon*," *American Art Journal* 13, no. 1 (1981): 47–68; "Peale's Mammoth," in Laura Rigal, *The American Manufactory: Art, Labor, and the World of Things in the Early Republic* (Princeton, NJ: Princeton University Press, 1998), 91–113; and "The Great American 'Incognitum,'" in Charles Coleman Sellers, *Mr. Peale's Museum: Charles Willson Peale and the First Popular Museum of Natural Science and Art* (New York: W. W. Norton, 1980), 123–58.

7. The vast literature on the history of European collecting includes Lorraine Daston and Katharine Park, *Wonders and the Order of Nature, 1150–1750* (New York: Zone Books, 1998); Oliver Impey and Arthur MacGregor, eds., *The Origins of Museums: The Cabinet of Curiosities in Sixteenth- and Seventeenth-Century Europe* (New York: Oxford University Press, 1985); Joy Kenseth, ed., *The Age of the Marvelous* (Hanover, NH: Hood Museum of Art, Dartmouth College, 1991); and Arthur MacGregor, *Curiosity and Enlightenment: Collectors and Collections from the Sixteenth to the Nineteenth Century* (New Haven, CT: Yale University Press, 2007).

8. Quoted in David R. Brigham, *Public Culture in the Early Republic: Peale's Museum and Its Audience* (Washington, DC: Smithsonian Institution Press, 1995), 5.

9. Peale wanted to purge American science of the theorizing and generalizing inaccuracies of Georges-Louis Leclerc, the count of Buffon, and his European contemporaries, and he called for a science based on observed facts and eyewitness testimony. Claiming that the visitors to his museum would be confused by randomly displayed material, he maintained that it was "only the arrangement and management of a Repository of Subjects of Natural History, that can constitute its utility" and, unless "the proper modes of seeing and using them [are] attended to, the advantage of such a store will be of little account to the public" (quoted in Stephen T. Asma, *Stuffed Animals and Pickled Heads: The Culture and Evolution of Natural History Museums* [New York: Oxford University Press, 2001], 170). On the organization of the collections of Peale's museum, see Roger B. Stein, "Charles Willson Peale's Expressive Design: *The Artist in His Museum*," in Marianne Doezema and Elizabeth Milroy, eds., *Reading American Art* (New Haven, CT: Yale University Press, 1998), 38–78. On Peale's wooden forms and glass eyes, see Robert McCracken Peck, "Preserving Nature for Study and Display," in Sue Ann Prince, ed., *Stuffing Birds, Pressing Plants, Shaping Knowledge: Natural History in North America, 1730–1860* (Philadelphia: American Philosophical Society, 2003), 11–25. The display of birds occupies a significant portion of Peale's self-portrait not only because birds formed an important part of the museum's collections but also because Peale believed they offered pertinent instruction for the citizens of the United States. Rising from specimens of ducks and penguins in the bottom cabinets, through songbirds at the middle and birds of prey in the top row, the avian display showed examples of patience, parental devotion, industriousness, and other virtues that Peale wished fellow citizens to cultivate.

10. On Peale's desire to embalm the Founding Fathers, see Asma, *Stuffed Animals and Pickled Heads*, 55.

11. Quoted in Sidney Hart, " 'To Encrease the Comforts of Life': Charles Willson Peale and the Mechanical Arts," in Lillian B. Miller and David C. Ward, eds., *New Perspectives on Charles Willson Peale* (Pittsburgh: University of Pennsylvania Press for Smithsonian Institution, 1991), 257–58.

12. Thomas Jefferson, *Notes on the State of Virginia*, 1787 (rpt., New York: Penguin Classics, 1988), 167. The significance of the fact that the mastodon had gone extinct confounded many Americans; Jefferson refused to believe it and maintained that mastodons still roamed the far reaches of the West.

13. On the history of perspective in painting and the interaction of the visual arts with the science of optics, see Martin Kemp, *The Science of Art: Optical Themes in Western Art from Brunelleschi to Seurat* (New Haven, CT: Yale University Press, 1990).

14. Peale summed up his life's work and the purpose of his museum in 1819 when he wrote to Thomas Jefferson, "The attainment of Happiness, Individual as well as Public, depends on the cultivation of the human mind" (quoted in Miller, "Charles Willson Peale as History Painter," 67). On the divisions and similarities between artistic and scientific creativity, see Jacob W. Getzels and Mihaly Csikszentmihalyi, "Scientific Creativity," *Science Journal* 3, no. 9 (September 1967): 80–84.

15. Many historians have written about the popularization of science in the United States during the nineteenth century. Some of the excellent sources on this subject include Toby A. Appel, "Science, Popular Culture, and Profit: Peale's Philadelphia Museum," *Journal of the Society for the Bibliography of Natural History* 9, no. 4 (April 1980): 619–34; John Rickards Betts, "P. T. Barnum and the Popularization of Natural History," *Journal of the History of Ideas* 20, no. 3 (June–September 1959): 353–68; Sally Gregory Kohlstedt, "Parlors, Primers, and Public Schooling: Education for Science in Nineteenth-Century America," *Isis* 81, no. 3 (September 1990): 424–45; Hyman Kuritz, "The Popularization of Science in Nineteenth-Century America," *History of Education Quarterly* 21, no. 3 (Fall 1981): 259–74; Margaret W. Rossiter, "Benjamin Silliman and the Lowell Institute: The Popularization of Science in Nineteenth-Century America," *New England Quarterly* 44, no. 4 (December 1971): 602–26; and Donald Zochert, "Science and the Common Man in Ante-Bellum America," *Isis* 65, no. 4 (December 1974): 448–73.

16. Quoted in I. Bernard Cohen, *Some Early Tools of American Science* (Cambridge, MA: Harvard University Press, 1950), 90. Some of my descriptions of *Man of Science* were first published in my article "The Great American Experiment," *American Art* 23, no. 2 (Summer 2009): 16–20.

17. Quoted in John Michael Vlach, *Plain Painters: Making Sense of American Folk Art* (Washington, DC: Smithsonian Institution Press, 1988), 57.

18. Quoted in Sally Gregory Kohlstedt, "In from the Periphery: American Women in Science, 1830–1880," *Signs: Journal of Women in Culture and Society* 4, no. 1 (August 1978): 90. On the gender divisions of the painting, see also Rigal, *American Manufactory*, 111–12. On nineteenth-century American women in science and, more generally, on gender and science, see Lois Barber Arnold, *Four Lives in Science: Women's Education in the Nineteenth Century* (New York: Schocken Books, 1984); Evelyn Fox Keller, *Reflections on Gender and Science* (New Haven, CT: Yale University Press, 1985); Judith A. McGaw, "Women Inventors, Engineers, Scientists, and Entrepreneurs," in Martha Moore Trescott, ed., *Dynamos and Virgins Revisited: Women and Technological Change in History—An Anthology* (Metuchen, NJ: Scarecrow Press, 1979), 100–179; Margaret W. Rossiter, *Women Scientists in America: Struggles and Strategies to 1940* (Baltimore: Johns Hopkins University Press, 1982); and Deborah Jean Warner, "Science Education for Women in Antebellum America," *Isis* 69, no. 1 (March 1978): 58–67.

19. Quoted in Pnina G. Abir-Am and Dorinda Outram, eds., *Uneasy Careers and Intimate Lives: Women in Science, 1789–1979* (New Brunswick, NJ: Rutgers University Press, 1987), n.35, 304.

20. American women made valuable contributions in geology, zoology, and especially botany, but they remained frustrated by the unwillingness of male counterparts to recognize their work or admit them to positions of leadership. The story of Elizabeth Britton, a leading nineteenth-century botanist, is telling. After many years of study and numerous publications and discoveries, Britton was conceded the title of Honorary Curator of Mosses at the New York Botanical Garden.

21. Quoted in Deborah Jean Warner, "Women Astronomers," *Natural History* 58, no. 5 (May 1979): 14.

22. Helen Wright, *Sweeper in the Sky: The Life of Maria Mitchell, First Woman Astronomer in America* (Clinton Corners, NY: College Avenue Press, 1997), 64. On Maria Mitchell's pursuit of astronomy, see Sally Gregory Kohlstedt, "Maria Mitchell and the Advancement of Women in Science," in Abir-Am and Outram, *Uneasy Careers and Intimate Lives*, 129–46. In the 1860s a British female author pondered the small number of women who studied advanced science or worked as professional scientists. She wrote: "The fact is indisputable that at the present time the students of science among men greatly outnumber those among women. Some persons attribute this circumstance to an inherent specific distinction in the minds of the two sexes of man.... Others perceive in existing social and conventional arrangements which exclude women from those opportunities of cultivating their intellectual faculties.... The last I think is the true solution of the question, 'Why are there fewer scientific women than scientific men?'" Lydia Ernestine Becker, "On the Study of Science by Women," *Contemporary Review* 10 (January–April 1869): 387.

23. Abraham Lincoln, "Second Lecture on Discoveries and Inventions" (1859), in Roy P. Basler, ed., *The Collected Works of Abraham Lincoln* (New Brunswick, NJ: Rutgers University Press, 1953), 3:363.

24. Mark Twain, *A Connecticut Yankee in King Arthur's Court* (New York: Harper and Brothers, 1889), 64.

25. The U.S. Patent Law of 1790, the first of its kind in the world, decreed that patent holders retained the rights to their patents for fourteen years. Although only three patents were issued in the year the Patent Law was enacted, the numbers of applications and patents that were granted increased exponentially in the ensuing decades. From 1820 to 1829, 2,697 patents were issued; from 1850 to 1859 the number climbed to 21,302; and from 1860 to 1869, 77,355 patents were awarded. On the egalitarian nature of the U.S. patent system, see B. Zorina Khan, *The Democratization of Invention: Patents and Copyrights in American Economic Development, 1790–1920* (Cambridge: Cambridge University Press, 2005). In 1888 the U.S. Patent Office began to take separate notice of female inventors and published a roster, *Women Inventors to Whom Patents Have Been Granted by the United States Government, 1790 to July 1, 1888*. Only fifty-five women patentees were listed before 1860, but by the middle of 1888 some three thousand more names had been added. The author of the women's patent list remarked that many of the patents issued in previous decades were registered by women using initials in place of their first names; assumed to be males, they were not counted as female patentees. "A slight investigation proves that patents taken out in some man's name are, in many instances, due to women," she noted. Matilda Joslyn Gage, "Woman as an Inventor," *North American Review* 136, no. 318 (May 1883): 483.

26. Quoted in Paul J. Staiti, *Samuel F. B. Morse* (Cambridge: Cambridge University Press, 1989), 33.

27. As recounted by Asher B. Durand in 1855. Quoted in John W. McCoubrey, ed., *American Art, 1700–1960: Sources and Documents* (Englewood Cliffs, NJ: Prentice-Hall, 1965), 113.

28. *Harper's New Monthly Magazine*, June 1859, 5.

29. Samuel F. B. Morse to James Fenimore Cooper, 1849. Quoted in Staiti, *Samuel F. B. Morse*, 208.

30. Quoted in Robert Luther Thompson, *Wiring a Continent: The History of the Telegraph Industry in the United States, 1832–1866* (Princeton, NJ: Princeton University Press, 1947), 253.

31. *Yankee Doodle*, October 1846, 1.

32. Quoted in Peter C. Marzio, *The Art Crusade: An Analysis of American Drawing Manuals, 1820–1860* (Washington, DC: Smithsonian Institution Press, 1976), 22. In a study of creativity published in the 1960s, Jacob Getzels and Mihaly Csikszentmihalyi also commented on the false division of art and science: "The artificially sustained divisions and resumed incompatibility between 'art' and 'science,' the one 'intuitive' and the other 'rational,' are not only gratuitous but actively antithetical to creativity. Creative achievement in science as elsewhere—indeed, the fullest realization of the individual—depends upon a union of the intuitive and the rational, of the imaginative and the analytic, of fantasy and control: a construction of the divergent and convergent aspects of thought." Getzels and Csikszentmihalyi, "Scientific Creativity," 84.

DEMOCRATIC TIME

33. John Adams described the synchronization of the American Revolution in a letter of February 13, 1818, to Hezekiah Niles. "Thirteen clocks were made to strike together," he wrote, "a perfection of mechanism, which no artist had ever before effected." Charles Francis Adams, ed., *The Works of John Adams, Second President of the United States* (Boston: Little, Brown, 1850–56), 10:283.

34. Michael Chevalier, *Society, Manners, and Politics in the United States*, trans. from 3rd Paris ed. (Boston: Weeks, Jordan, 1839), 309. Some of the ideas in this chapter were first published in Perry, "Great American Experiment," 16–20.

35. J. N. Bellows, "An Explanation of American Characteristics," *Hunt's Merchants' Magazine* 8 (1843): 167–68. On nineteenth-century Americans' ideas about time, see Thomas M. Allen, *A Republic in Time: Temporality and Social Imagination in Nineteenth-Century America* (Chapel Hill: University of North Carolina Press, 2008); Ian R. Bartky, *Selling the True Time: Nineteenth-Century Timekeeping in America* (Stanford, CA: Stanford University Press, 2000); Michael O'Malley, *Keeping Watch: A History of American Time* (New York: Viking, 1990); and Carlene E. Stephens, *On Time: How America Has Learned to Live by the Clock* (Boston: Bullfinch, 2002).

36. On peddlers of Terry's clocks, see David Jaffee, "Peddlers and Progress: The Transformation of the Rural North, 1760–1860," *Journal of American History* 78, no. 2 (September 1991): 517.

37. Quoted in Roger Burlingame, *Machines That Built America* (New York: Signet Books, 1953), 9. Terry died in 1852. In the last years of his life, he had been selling approximately eleven thousand clocks a year for five dollars each. With his passing, other clock companies rushed into the market to satisfy Americans' hunger for timekeeping devices. The clock industry became an attractive business for speculators, although many failed to invest their funds with the right clockmaker at the right time. One investor to lose his shirt in clock investments was P. T. Barnum, who was forced to cover his debts by selling the natural history collections that he had bought years before from the Philadelphia museum of Charles Willson Peale.

38. Quoted in David E. Nye, *America as Second Creation: Technology and Narratives of New Beginnings* (Cambridge, MA: MIT Press, 2003), 18.

39. For additional information on the Van Dame portrait, see Nancy B. Tieken et al., *American Art from the Currier Gallery of Art* (New York: American Federation of Arts, 1995), 33–34.

40. Harriet Beecher Stowe, *The Minister's Wooing* (Boston: Brown, Taggard, and Chase, 1859), 2–3.

41. Catharine Beecher, *Treatise on Domestic Economy, for the Use of Young Ladies at Home, and at School* (Boston: Marsh, Capen, Lyon, and Webb, 1841), 175. On the busy-ness of nineteenth-century American women, see also my article, "American Women, Neurasthenia, and the Art of Doing Nothing," in Katherine Williams et al., *Women on the Verge: The Culture of Neurasthenia in Nineteenth-Century America*, exh. cat. (Stanford, CA: Iris and B. Gerald Cantor Center for Visual Arts, 2004): 69–77.

42. On the difference between female time and male time, see Nancy F. Cott, *The Bonds of Womanhood: "Woman's Sphere" in New England, 1780–1835* (New Haven, CT: Yale University Press, 1977), 60–62.

43. Women's inventions related predominantly to women's household tasks, such as preserving fruits and vegetables, cooking, cleaning, and making and washing clothes. Despite the general lack of recognition for female inventors, innovations by women found a ready market, and the household economy remained a center of creative tinkering by females. The nineteenth-century activist and author Matilda Joslyn Gage commented on American society's failure to acknowledge the inventions made by nineteenth-century American women. "Deprived, as a woman is, of political power, she has to face contempt of her sex, open and covert scorn of womanhood, depreciatory allusions to her intellectual powers—all tending to hamper the expression of her inventive genius" (Gage, "Woman as an Inventor," 488). On American women who patented their timesaving inventions, see Khan, *Democratization of Invention*, 132–45.

44. Quoted in Alice Kessler-Harris, *Out to Work: A History of Wage-Earning Women in the United States* (New York: Oxford University Press, 2003), 23–24. Until immigrants filled the gap at midcentury, native-born female operatives made up a significant part of factory labor in the United States. Men sought jobs of higher status and pay or worked on farmland they owned. On female factory operatives and women as wage earners, see Thomas Dublin, *Transforming Women's Work: New England Lives in the Industrial Revolution* (Ithaca, NY: Cornell University Press, 1994); and "Industrial Wage Earners and the Domestic Ideology," in Kessler-Harris, *Out to Work*, 45–72.

45. *Factory Girls' Garland*, May 25, 1844. Quoted in Philip S. Foner, ed., *The Factory Girls: A Collection of Writings on Life and Struggles in the New England Factories of the 1840's* (Urbana: University of Illinois Press, 1977), 77. In the mills, girls who worked twelve- to fourteen-hour days were expected to avail themselves of night-school courses and self-improvement circles that mill owners encouraged to maintain their reputations as benevolent employers. Female mill operatives were among the leaders of the nineteenth-century movement for the ten-hour workday.

46. The painting was exhibited at the Century Association, New York, in 1871 as "The Old Mill." For additional information on the painting, see Nicolai Cikovsky Jr., "Winslow Homer's (So-called) *Morning Bell*," *American Art Journal* 29, nos. 1 and 2 (1998): 4–17; and Bryan J. Wolf, "The Labor of Seeing: Pragmatism, Ideology, and Gender in Winslow Homer's *The Morning Bell*," *Prospects* 17 (1992): 273–318.

47. As a gang manager at the Midvale Steel Works, Taylor began to consider how long industrial tasks took and how long they should take. He started to think of the flow of movements in segments, breaking them into distinct components. Analyzing the

work that went into making locomotive tires, for example, Taylor discovered a train of elements: setting the tire on the machine, roughing out the face of the front edge, finishing the face's front edge, rough boring the front, and finish boring the front. He assigned ideal times for the completion of each task and noted them on what he termed an "instruction card." The card, which sometimes ran to several pages, listed the tools needed for a job, as well as the order in which each operation was to be performed (Robert Kanigel, *The One Best Way: Frederick Winslow Taylor and the Enigma of Efficiency* [New York: Viking, 1997], 206–7). For additional information on Taylor, see Jill Lepore's essay on scientific management, "Not So Fast," *New Yorker*, October 12, 2009, 84–122; and Daniel Nelson, *Frederick W. Taylor and the Rise of Scientific Management* (Madison: University of Wisconsin Press, 1980).

48. On Eadweard Muybridge's locomotion studies, see Phillip Prodger, *Time Stands Still: Muybridge and the Instantaneous Photography Movement*, exh. cat. (New York: Oxford University Press for Cantor Center for Visual Arts, Stanford University, 2003).

49. Quoted in Wolf, "Labor of Seeing," 278. It is noteworthy that at his celebrated research complex in Menlo Park, the famed inventor Thomas Edison had the springs removed from every clock in every laboratory to remind employees that inventive activities could not be ruled by the clock. Edison, who often labored for twenty-four-hour stretches interspaced with short naps, maintained a fluid work schedule there.

50. The huge variety of paper currency issued by American banks made paper money a charged subject in nineteenth-century life. After the Civil War, currency became an urgent problem owing to the spread of worthless Confederate money and widespread counterfeiting. In 1865 an estimated 40 percent of the currency in circulation in the United States was counterfeit. On the depiction of money in nineteenth-century American art, see Bruce W. Chambers, *Old Money: American Trompe L'Oeil Images of Currency* (New York: Berry-Hill Galleries, 1988); and Edward Nygren, "The Almighty Dollar: Money as a Theme in American Painting," *Winterthur Portfolio* 23, nos. 2 and 3 (Summer/Autumn 1988): 129–50.

51. Quoted in Michael Chevalier, *Society, Manners, and Politics in the United States: Letters on North America*, ed. John W. Ward, trans. T. G. Bradford (Gloucester, MA: Peter Smith, 1967), 271.

52. Quoted in Stephen Van Dulken, *Inventing the Nineteenth Century: One Hundred Inventions That Shaped the Victorian Age from Aspirin to the Zeppelin* (New York: New York University Press, 2001), 72.

NIAGARA FALLS

53. "Niagara Seen with Different Eyes," *Harper's Weekly*, August 9, 1873, 698.

54. For additional information on Barralet, see Elizabeth McKinsey, "An American Icon," in Jeremy Elwell Adamson, *Niagara: Two Centuries of Changing Attitudes, 1697–1901*, exh. cat. (Washington, DC: Corcoran Gallery of Art, 1985), 87.

55. Quoted in Elizabeth McKinsey, *Niagara Falls: Icon of the American Sublime* (Cambridge: Cambridge University Press, 1985), 49.

56. Wilson's *American Ornithology* was one of the earliest and most important of the natural history inventories created during the early nineteenth century. Its list of subscribers stretched to every corner of the country—New Hampshire, New York, Pennsylvania, Kentucky, South Carolina, Mississippi—testifying to the existence of a thriving national community of bird lovers. Wilson's subscribers represented many of the nation's most eminent citizens, among them, Thomas Jefferson, James Madison, Jedediah Morse, and Gouverneur Morris. Charles Willson Peale included in *The Artist in His Museum* (fig. 2, p. 5) a picture of the stuffed eagle that Wilson used as the model for the one in his book. The bird had been Peale's pet for many years, and the artist wrote in his autobiography that the eagle "had so long been domesticated, that Peale could without fear stroke him with his hand, nay it knew him so well that when [Peale] was walking in the State House Garden, it would utter cries expressive of its pleasure." Quoted in Edgar P. Richardson et al., *Charles Willson Peale and His World* (New York: Harry N. Abrams, 1983), 114.

57. Emma Willard, *Geography for Beginners* (Hartford, CT: Oliver D. Cooke, 1826), 10.

58. Quoted in Anne Baker, *Heartless Immensity: Literature, Culture, and Geography in Antebellum America* (Ann Arbor: University of Michigan Press, 2006), 24.

59. Quoted in Adamson, *Niagara*, 32.

60. The full title of Raphaelle Peale's painting is *Venus Rising from the Sea—A Deception (After the Bath)*. There is no evidence that Peale visited Niagara Falls or had it in mind when he painted *Venus Rising*. Yet in the context of the plethora of "behind the sheet" anecdotes and published accounts circulating in the United States at the time he created it, its composition and watery allusions make an intriguing connection to Niagara that may have been obvious to nineteenth-century American audiences. The full meaning of Peale's visual joke, however, remains mysterious; scholars Lauren Lessing and Mary Schafer, for example, do not address the possibility of a reference to Niagara in their discussion of the painting's meanings in "Unveiling Raphaelle Peale's *Venus Rising from the Sea—A Deception*," *Winterthur Portfolio* 43, nos. 2 and 3 (Summer/Autumn 2009): 229–59.

61. Frances Trollope, *Domestic Manners of the Americans*, 1832 (rpt., Donald Smalley, ed. [London: Folio Society, 1974]), 280.

62. Basil Hall, *Travels in North America in the Years 1827 and 1828* (Edinburgh: Cadell, 1829), 1:169.

63. Thomas Cole, "Essay on American Scenery," 1835. Quoted in McCoubrey, *American Art, 1700–1960*, 105.

64. J. L. Comstock, *Outlines of Geology* (Hartford, CT: D. F. Robinson, 1834), 79. On the subject of the relationship that nineteenth-century Americans saw between the study of geology and the study of scripture, see Herbert Hovenkamp's chapter, "A Theory of the Earth," in his *Science and Religion in America, 1800–1863* (Philadelphia: University of Pennsylvania Press, 1978), 119–45; and Conrad Wright, "The Religion of Geology," *New England Quarterly* 14, no. 2 (1941): 335–58.

65. On Cole's passion for geology, see Rebecca Bedell, "Thomas Cole and the Fashionable Science," *Huntington Library Quarterly: Studies in English and American History and Literature* 59, nos. 2 and 3 (1996): 348–78. On the broader subject of nineteenth-century American artists' interest in geology, see Bedell's excellent study *The Anatomy of Nature: Geology and American Landscape Painting, 1825–1875* (Princeton, NJ: Princeton University Press, 2001).

66. Before he had visited Niagara, Church copied Cole's *Distant View of Niagara Falls* (fig. 31, p. 58) in 1846. See McKinsey, *Niagara Falls*, 243.

67. Quoted in John Wilmerding, ed., *American Light: The Luminist Movement, 1850–1875*, exh. cat. (Washington, DC: National Gallery of Art, 1980), 67.

68. David Livingstone, *Missionary Travels and Researches in South Africa* (New York: Harper and Brothers, 1858), 558.

69. Quoted in Franklin Kelly, *Frederic Edwin Church*, exh. cat. (Washington, DC: National Gallery of Art and Smithsonian Institution Press, 1989), 52.

70. Quoted in McKinsey, *Niagara Falls*, 244.

71. On waterpower in the United States, see Louis Hunter, *A History of Industrial Power in the United States, 1780–1930*, vol. 1, *Waterpower in the Century of the Steam Engine* (Charlottesville: University Press of Virginia for Eleutherian Mills–Hagley Foundation, 1979); and "Water and Industry," in David E. Nye, *Consuming Power: A Social History of American Energies* (Cambridge, MA: MIT Press, 1998), 43–68. Kenneth L. Sokoloff, in his study of patent activity in the antebellum United States, found that during the nineteenth century patent applications were strongly associated with patent seekers' residence near rivers. See his article "Inventive Activity in Early Industrial America: Evidence from Patent Records, 1790–1846," *Journal of Economic History* 48, no. 4 (December 1988): 813–50.

72. Quoted in Linda L. Revie, *The Niagara Companion: Explorers, Artists, and Writers at the Falls, from Discovery through the Twentieth Century* (Waterloo, ON: Wilfrid Laurier University Press, 2003), 128. It is intriguing to consider Church's grand painting of American waterpower in the context of the severe economic crisis of 1857 that shuttered the doors of countless water-powered factories and decimated related businesses across the country. In the year that the painting made its debut, *Scientific American* reported that "no one but an eye-witness can have a conception of [manufacturers'] complete and overwhelming prostration. Factories are closed, forge fires are extinguished; the hammer, the saw, the spindle, and the loom are silent; and men walk about the streets with anxious, care-worn countenances." Quoted in "Fifty, One Hundred, and 150 Years Ago," *Scientific American*, November 2007, 14.

73. Hunter, *Waterpower*, 1:205; and Dirk J. Struik, *Yankee Science in the Making: Science and Engineering in New England from Colonial Times to the Civil War* (New York: Dover, 1948), 321–23. On the theme of the river in nineteenth-century American art, see "American Waters: The Flow of the Imagination," in John Wilmerding, *American Views: Essays on American Art* (Princeton, NJ: Princeton University Press, 1991), 49–68.

74. P. T. Barnum, *The Life of P. T. Barnum, Written by Himself*, 1855 (rpt., Urbana: University of Illinois Press, 2000), 226–27.

75. Quoted in Jeremy Elwell Adamson, *Niagara: Two Centuries of Changing Attitudes, 1697–1901*, exh. cat. (Washington, DC: Corcoran Gallery of Art, 1985), 75–76.

76. H. E. D., "The Fugitive Slave's Apostrophe to Niagara," *Boston Courier*, November 1, 1841. Quoted in Joseph T. Buckingham, *Personal Memoirs and Recollections of Editorial Life* (Boston: Ticknor, Reed, and Fields, 1852), 2:192.

77. On Benjamin Montgomery, see Portia P. James, *The Real McCoy: African-American Invention and Innovation, 1619–1930* (Washington, DC: Smithsonian Institution Press for Anacostia Museum, Smithsonian Institution, 1989), 52–53; and Patricia Carter Sluby, *The Inventive Spirit of African Americans: Patented Ingenuity* (Westport, CT: Praeger, 2004), 30–36. On the broader subject of nineteenth-century African American inventors, see also Daniel Murray, "Who Invented the Cotton Gin?" *Voice of the Negro* 2, no. 2 (February 1905): 96–102; and James C. Williams, comp., *At Last Recognition in America: A Reference Handbook of Unknown Black Inventors and Their Contributions to America* (Chicago: B.C.A., 1978).

78. Quoted in Revie, *Niagara Companion*, 143.

79. Quoted in Pierre Berton, *Niagara: A History of the Falls* (New York: Kodansha International, 1997), 163.

80. Quoted in Revie, *Niagara Companion*, 149.

THE PEACEMAKER

81. Quoted in Lee Kennett and James La Verne Anderson, *The Gun in America: The Origins of a National Dilemma* (Westport, CT: Greenwood Press, 1975), 42–43.

82. Quoted in Charles Royster, *A Revolutionary People at War: The Continental Army and American Character, 1775–1783* (Chapel Hill: University of North Carolina Press for Institute of Early American History and Culture, Williamsburg, Virginia, 1979), 34. Colonial inhabitants of the early frontier came up with firearm improvements to suit their circumstances, and German immigrant gunsmiths living in the region of Lancaster, Pennsylvania, made many useful gun innovations. They took the large-bore, short-barreled Jaeger rifle and made it a smaller caliber to conserve powder and lead and lengthened the barrel. The Pennsylvania long rifle was lighter and cheaper to use, and it employed a greased patch to seat the ball, which formerly needed to be hammered, to reduce loading time. Called the Pennsylvania or Kentucky rifle, it became the weapon of choice for most colonial Americans.

83. Quoted in Royster, *Revolutionary People at War*, 59.

84. The disinclination of German mercenaries to perish on behalf of the British monarch was also a significant factor.

85. *Analectic Magazine* (November 1815): 363–64. On the growth of a cult of marksmanship in the United States during the nineteenth century, see Kennett and Anderson, *Gun in America*.

86. On Edmonds's play on the colloquialism of "image peddling" as a term for prevalent practice of promoting political candidates with idealized images, see Elizabeth Johns, *American Genre Painting: The Politics of Everyday Life* (New Haven, CT: Yale University Press, 1991), 54.

87. The full sentence of the Second Amendment of the U.S. Constitution quoted here reads: "A well-regulated Militia, being necessary to the security of a free State, the right of the people to keep and bear Arms, shall not be infringed." Opposing interpretations surround this vaguely worded amendment. One view says that it intends to protect an individual's right to own, possess, and transport guns. Another says that the provision was not meant to protect individuals, but rather to shield states from federal legislation restricting their ability to maintain formal, organized militias. In 2008, a majority of Supreme Court justices endorsed the individual right to bear arms in *District of Columbia vs. Heller,* which struck down a District of Columbia ban on handguns. In their dissent, Justices Breyer, Stevens, Souter, and Ginsburg asserted that, while the Second Amendment affirmed an individual's right to bear arms as part of a "well-regulated militia," it did not confer on individuals an inalienable right to own a firearm. For an overview of the controversies surrounding individual gun ownership, see Gregg Lee Carter, ed., *Guns in American Society: An Encyclopedia of History, Politics, Culture, and the Law* (Santa Barbara, CA: ABC-CLIO, 2002), 2:525.

88. Quoted in Jon Meacham, *American Lion: Andrew Jackson in the White House* (New York: Random House, 2008), 25–30.

89. Quoted in Kennett and Anderson, *Gun in America,* 144.

90. Alexander Hamilton, on the night before his duel with Aaron Burr, penned four pages of what he called "remarks explanatory of my conduct, motives and views" to explain the code of honor that prompted him to accept Burr's challenge. "All considerations which constitute what men of the world denominate honor, impressed on me (as I thought) a peculiar necessity not to decline the call," he wrote. "The ability to be in future useful, whether in resisting mischief or effecting good, in those crises of our public affairs, which seem likely to happen, would probably have been inseparable from a conformity with public prejudice in this matter" (quoted in Joanne B. Freeman, "Dueling as Politics: Reinterpreting the Burr-Hamilton Duel," *William and Mary Quarterly,* 3rd ser., 53, no. 2 [April 1996]: 291–92). On the subject of dueling in the early nineteenth century, see also Joyce Appleby, *Inheriting the Revolution: The First Generation of Americans* (Cambridge, MA: Belknap Press of Harvard University Press, 2001), 41–45.

91. Both newspapers are quoted in Philip D. Jordan, *Frontier Law and Order: Ten Essays* (Lincoln: University of Nebraska Press, 1970), 8.

92. Quoted in ibid., 38–39.

93. Quoted in Edward Pessen, *Jacksonian America: Society, Personality, and Politics.* Rev. ed. (Urbana: University of Illinois Press, 1978), 37.

94. Quoted in Ron Tyler, *American Frontier Life: Early Western Paintings and Prints* (New York: Abbeville Press, 1987), 51–52.

95. Quoted in James H. Nottage, "Fashioning the West: The Pre-Twentieth-Century Origins of Western Wear," in Holly George-Warren and Michelle Freedman, *How the West Was Worn* (New York: Harry N. Abrams in association with Autry Museum of Western Heritage, Los Angeles, 2001), 14–15.

96. On the power of the frontier myth and the role of the gun in the American national narrative, see Richard Slotkin's exhaustive studies on the subject, *The Fatal Environment: The Myth of the Frontier in the Age of Industrialization, 1800–1890* (Middletown, CT: Wesleyan University Press, 1986), and *Regeneration through Violence: The Mythology of the American Frontier, 1600–1860* (Middletown, CT: Wesleyan University Press, 1973).

97. Quoted in Axelrod, *American Frontier Life,* 60.

98. Quoted in Johns, *American Genre Painting,* 70. Deas had applied to West Point but was not admitted. Raised in Philadelphia and New York, he sketched Indians in the West from 1840 to 1844 after seeing and being inspired by George Catlin's Indian Gallery (housed in the Smithsonian American Art Museum, Washington, DC; for more on the Indian Gallery, see chapter 4, "The Buffalo," in this book and n.126, below). Deas took to dressing like a trapper and began to use the name "Rocky Mountain." In St. Louis in 1847 he suffered a mental breakdown and stopped painting.

99. Quoted in Charles G. Worman, *Gunsmoke and Saddle Leather: Firearms in the Nineteenth-Century American West* (Albuquerque: University of New Mexico Press, 2005), 1. Before the 1850s these weapons were usually single-shot muzzle-loading rifles.

100. Quoted in Kennett and Anderson, *Gun in America,* 123.

101. Quoted in Axelrod, *American Frontier Life,* 94, which also includes the story of Joe Meek, the mountain man who claimed to be the subject of Ranney's painting. Meek and a party of trappers at the headwaters of the Missouri in 1837 were in battle for two days with Blackfeet; he reported that, with his last shot, he killed a Blackfoot about to ambush him and rode to safety. Meek called his gun "Sally."

102. Quoted in Carolyn C. Cooper, "A Connecticut Yankee Courts the World," in Herbert G. Houze, *Samuel Colt: Arms, Art, and Invention*, ed. by Elizabeth Mankin Kornhauser (New Haven, CT: Yale University Press and Wadsworth Atheneum Museum of Art, 2006), 1.

103. Quoted in William Hosley, *Colt: The Making of an American Legend* (Amherst: University of Massachusetts Press, 1996), 62. Early guns required hand finishing by a range of skilled workers, and their many components needed to be laboriously adjusted to fit into one another. Colt revolvers were not manufactured with exchangeable parts; they were standardized, but workmen needed to file and fit them. When finished, all the parts were stamped with a serial number as the metal hardened. "Including screws, there are twenty eight pieces in a pistol, and three times as many in a regulation musket.... Their value is apparent only when they are fitly joined together in one organic whole compacted of many parts, all mutually related—E pluribus Unum." Henry Barnard, *Armsmear: The Home, the Arm, and the Armory of Samuel Colt: A Memorial*, 1866 (rpt., Maryland: Beinfeld, 1976), 242.

104. Mark Twain, "A Glimpse of Hartford," *Alta California*, March 3, 1868, 2. The "American system" of Colt's factory was pioneered by numerous innovators, including Eli Whitney, owner of the patent for the cotton gin. Improvements in the armory system replaced the painstaking work of a gunsmith, who in those days was called an "artist." Even a smoothbore musket was a work of art almost like a painting or a piece of sculpture, and firearm experts could spot small differences that indicated the individual makers, many of whom were famous. Colt's factory represented the fruition of decades of incremental improvements in the small-arms industry in the United States, most of it occurring in the national armories in Springfield, Massachusetts, and Harpers Ferry, West Virginia. The work of army ordnance officers, inventive machinists, and private arms contractors, the innovations in manufacturing methods for interchangeable parts achieved in Springfield and Harpers Ferry spread to factories making clocks, sewing machines, railroad parts, hand-tools, farm implements, shoes, and wagons.

105. Colt made arms that were neither for militia training nor for bagging game, even though Catlin helped advertise them as hunting weapons. They were small arms that were easily concealable and ideal for violent encounters of both gold rush and settled populations. The price of Colt's weapons was within the reach of average people, and because his revolvers fired multiple rounds, the skill of the handler was not paramount for successful shooting. Colt fabricated guns for every taste, including forty-four

styles of pistols in six patterns, eleven lengths, and twenty-seven finishes. Jan E. Dizard, Robert Merrill Muth, and Stephen P. Andrews Jr., eds., *Guns in America: A Reader* (New York: New York University Press, 1999), 83.

106. Michael A. Bellesiles, *Arming America: The Origins of a National Gun Culture* (New York: Alfred A. Knopf, 2000), 431.

107. Quoted in ibid., 426.

108. Hart, " 'To Encrease the Comforts of Life,' " 239.

109. Jeff Cooper, *The Pistol as a Weapon of Defence in the House and on the Road: How to Choose It and How to Use It*, 1875 (rpt., New York: Paladin Press, 2004), 13.

110. Quoted in Drew Gilpin Faust, *This Republic of Suffering: Death and the American Civil War* (New York: Alfred A. Knopf, 2008), 42.

111. Quoted in Wolf, "Labor of Seeing," 312.

112. On the Berdan Sharpshooters, Lincoln, and Civil War ordnance improvements made under Lincoln's direction, see Robert V. Bruce, *Lincoln and the Tools of War* (Indianapolis, IN: Bobbs-Merrill, 1956), 107–17.

113. Quoted in Julia Keller, *Mr. Gatling's Terrible Marvel: The Gun That Changed Everything and the Misunderstood Genius Who Invented It* (New York: Viking, 2008), 42–43.

114. On the estimates of the number of men killed in the Civil War, see James McPherson, *Battle Cry of Freedom: The Civil War Era* (New York: Oxford University Press, 1988), 306n.

115. Kennett and Anderson, *Gun in America*, 93.

116. Elisha J. Lewis, M.D., *The American Sportsman: Containing Hints to Sportsmen, Notes on Shooting* (Philadelphia: J. B. Lippincott, 1863), xiii.

117. Ibid., 49.

118. "Didymus" (Martin Johnson Heade), *Forest and Stream* 14, no. 7 (March 18, 1880): 132.

119. Ibid. On hunters as conservationists during the nineteenth century, see "Conservation and Conflict," in Daniel Justin Herman, *Hunting and the American Imagination* (Washington, DC: Smithsonian Institution Press, 2001), 237–53.

120. "Breech-loading Fire-arms," in Horace Greeley et al., *The Great Industries of the United States: Being a Historical Summary of the Origin, Growth, and Perfection of the Chief Industrial Arts of This Country* (Hartford, CT: J. B. Burr and Hyde, 1872), 812.

THE BUFFALO

121. Quoted in Godman, *American Natural History*, pt. 3, 2.

122. Quoted in John C. Greene, "American Science Comes of Age, 1780–1820," *Journal of American History* 55, no. 1 (June 1968): 35.

123. Quoted in Lee Alan Dugatkin, *Mr. Jefferson and the Giant Moose: Natural History in Early America* (Chicago: University of Chicago Press, 2009), 23.

124. The term "bison" is the correct name for the animal, but nineteenth-century Americans used that name interchangeably with "buffalo," as I have in this chapter. "Buffalo" may have originated with the English. In the 1500s, English military men wore protective jackets known as a buff coats, which were soft and thick and made of undyed leather. When they arrived in the New World, they described any animal—from moose to manatee—that yielded such leather as a "buff." The name eventually stuck with the bison, which had migrated to North America during the Pleistocene period with woolly mammoths, the American mastodon, and the saber-toothed tiger. For additional information on the etiology of the word "buffalo," see Steven Rinella, *American Buffalo: In Search of a Lost Icon* (New York: Spiegel and Grau, 2009), 41.

125. George Catlin, *Letters and Notes on the Manners, Customs, and Condition of the North American Indians*, 1841 (rpt., New York: Dover, 1973), 1:24.

126. Catlin decided to make his life's work an artistic reconnaissance of the tribes of North America and its topography, botany, and zoology. Beginning in 1830, he made five trips to the western frontier, studying and making paintings of North American tribes and their activities and environment. In 1837 he assembled his works and began public exhibitions of his Indian Gallery, which included many images of buffalo on the prairie. In 1841 he published a narrative of his travels that included images of buffalo and descriptions of their behavior and habitat.

127. Quoted in Andrew C. Isenberg, *The Destruction of the Bison: An Environmental History, 1750–1920* (Cambridge: Cambridge University Press, 2000), 23. Isenberg suggests that the plains may have supported twenty-seven million bison. On nineteenth-century Americans' reactions to emerging scientific evidence of species extinction, see Mark V. Barrow Jr., *Nature's Ghosts: Confronting Extinction from the Age of Jefferson to the Age of Ecology* (Chicago: University of Chicago Press, 2009).

128 Catlin, *Letters and Notes on North American Indians*, 1:247.

129 Ibid., 1:254.

130. On the transition of certain Native American tribes from planters and hunter-gathers to bison hunters, see William Cronon's chapter "Commodities of the Hunt," in his *Changes in the Land: Indians, Colonists, and the Ecology of New England* (New York: Hill and Wang, 1983), 82–107; Dan Flores, "Bison Ecology and Bison Diplomacy: The Southern Plains from 1800 to 1850," *Journal of American History* 78, no. 2 (September 1991): 465–85; and Andrew C. Isenberg, "Toward a Policy of Destruction: Buffaloes, Law, and the Market, 1803–1883," *Great Plains Quarterly* 12, no. 4 (Fall 1992): 227–41. Isenberg reports that by the 1840s, western plains nomads were bringing annually to western steamboats more than one hundred thousand bison robes, in addition to their personal take of roughly five hundred thousand bison yearly for subsistence and intertribal trade.

131. Catlin, *Letters and Notes on North American Indians*, 1:26–27.

132. Barnum, *Life of P. T. Barnum*, 253–24. On American sportsmen hunting the buffalo, see Herman, *Hunting and American Imagination*, 200–205.

133. On the effect of the refrigerator car in the development of western cattle ranching, see William Cronon, *Nature's Metropolis: Chicago and the Great West* (New York: W. W. Norton, 1991), 235–47. For an account of the men who hunted buffalo en masse for hides for industrial belting, see Rinella, *American Buffalo*, 168–83; and Mari Sandoz, *The Buffalo Hunters: The Story of the Hide Men* (New York: Hastings House, 1954).

134. Isenberg, *Destruction of the Bison*, 31.

135. The construction of western railroads had a significant effect on the American bison. The transcontinental railroad began in 1865, and with its completion in 1869 the buffalo were divided into two great herds that were never again united: a northern herd of about a million and a half animals and a southern herd of approximately five million. The northern herd ranged through the Powder River country and into British possessions. In the Southwest the buffalo were abundant in western Texas, but the favorite feeding ground of the southern herd comprised the sections of the plains near the Republican River, between what was then Arkansas and the South Platte.

176. Quoted in Huth, *Nature and the American*, 6. On early colonial ideas about the American wilderness, see also Amy Meyers, "Imposing Order on the Wilderness: Natural History Illustration and Landscape Portrayal," in Edward J. Nygren, *Views and Visions: American Landscape before 1830*, exh. cat. (Washington, DC: Corcoran Gallery of Art, 1986), 104–31. In 1857 a writer quoted in the *Crayon* associated "cruelty to trees" with fear of Indians. "The American seems to have a hereditary antipathy to Indians and trees," he wrote. "When every tree trunk might be the ambush of a skulking enemy whose war-whoop made the hair stand on end as a convenient handle to the insecure scalp (so does Nature adapt means to ends), a man might be pardoned for wishing to see as much cleared ground about his homestead as possible." James Russell Lowell, "Humanity to Trees," *Crayon* 4 (1857): 96.

177. Quoted in Cox et al., *This Well-Wooded Land*, 55–56.

178. On Harvey, see Roberta Owen, "George Harvey's Anglo-American Atmospheric Landscapes," *Magazine Antiques* 174, no. 4 (October 2009): 112–21. Harvey's images convey what British visitor Isaac Weld observed during his tour of the United States: "The man that can cut down the largest number of [trees], and have fields about his house is looked upon as the most industrious citizen, and the one that is making the greatest improvement in the country." The artist Thomas Cole had a very different view of Americans' forest-clearing work. In 1835, Cole described the consequences of the accelerating tempo of logging in the United States. "The ravages of the axe are daily increasing—the most noble scenes are made desolate, and oftentimes with a wantonness and barbarism scarcely credible in a civilized nation," he lamented. "The wayside is becoming shadeless, and another generation will behold spots, now rife with beauty, desecrated by what is called improvement" (Thomas Cole, "Essay on American Scenery," 1835, quoted in Harold Spencer, ed., *American Art: Readings from the Colonial Era to the Present* [New York: Charles Scribner, 1980], 89). On nineteenth-century portrayals of logging, see Nicolai Cikovsky Jr., " 'The Ravages of the Axe': The Meaning of the Tree Stump in Nineteenth-Century American Art," *Art Bulletin* 61, no. 4 (December 1979): 611–26; and Barbara Novak, "The Double-Edged Axe," *Art in America* 64 (January–February 1976): 44–50.

179. On nineteenth-century American ideas about forest clearing, see Cox et al., *This Well-Wooded Land*, 133–53.

180. Quoted in ibid., 55. The destruction of forests was also linked to the decimation of buffalo. The tanning industry that processed the hides for industrial belting required trees and lime in enormous quantities. The tanning industry initially concentrated in northeastern Pennsylvania and the Adirondack region of New York amidst dense forests of eastern hemlock, the bark of which is rich in tannin. The chemical is essential to the tanning process, and heavy harvesting of hemlock rapidly depleted the species. Slow-growing trees that take three hundred years to reach maturity, hemlocks as saplings rely on the shade of adjacent trees. By the 1880s the leather industry's prodigious use of tanbark had eradicated easily accessible supplies of hemlock; tanneries thus were part of a link in a system that depleted both bison herds and forests.

181. Quoted in Tamara Plakins Thorton, *Cultivating Gentlemen: The Meaning of Country Life among the Boston Elite, 1785–1860* (New Haven, CT: Yale University Press, 1989), 165. On the moral aspect of horticulture in the United States, see Thornton's "Cultivating the American Character: Horticulture and Moral Reform in Antebellum America," *Orion Nature Quarterly* 1, no. 4 (Spring 1985): 10–19. Some scholars have speculated that Rubens Peale's geranium was the first brought to America. Rubens was Charles Willson Peale's fourth son; as a child, he studied botany and mineralogy and worked in his father's museum. Rubens's interest in rare plants and seeds may have emerged because bad eyesight kept him from pursuing art. He gathered and germinated seeds for his father and successfully cultivated the delicate geranium, a plant that Thomas Jefferson had problems growing at the White House. For additional information on the portrait, see John Wilmerding's "Rembrandt Peale's Rubens Peale with a Geranium," in his *American Views: Essays on American Art* (Princeton, NJ: Princeton University Press, 1991), 143–57.

182. Cole, "Essay on American Scenery," quoted in McCoubrey, *American Art, 1700–1960*, 98–109.

183. Harland Coultas, *What May Be Learned from a Tree* (New York: D. Appleton, 1860), 24–25.

184. Quoted in David B. Lawall, "Asher Brown Durand: His Art and Art Theory in Relation to His Times" (Ph.D. diss., Princeton University, 1966), 372.

185. Quoted in Kruska, *Sierra Nevada Big Trees*, 22.

186. Ibid., 30–31.

187. Horace Greeley, *An Overland Journey from New York to San Francisco in the Summer of 1859* (San Francisco: H. H. Bancroft, 1860), 264. Big trees are not only the world's largest living things, but they are also estimated to be the fastest growing, not just in height but in overall volume. They increase at approximately an

inch in diameter per year, even when the trunk has reached one hundred feet in diameter. Furthermore, old age doesn't seem to exist for the trees, in that its life processes don't appear to slow down or change in a basic way. Death usually comes by fire, wind, or the erosion of soil. The oldest known specimen is three thousand years old, not as old as the bristlecone pine, but it is still growing. These remarkable trees also support a diverse community of plants and animals, including orchids, ferns, lilies, owls, eagles, fox, bears, puma, salamander, rattlesnakes, jays, woodpeckers, and kingfishers.

188. Quoted in Maria Morris Hambourg, "Carleton Watkins: An Introduction," in Douglas R. Nickel, *Carleton Watkins: The Art of Perception*, exh. cat. (San Francisco: San Francisco Museum of Modern Art; New York: Harry N. Abrams, 1999), 10.

189. Quoted in Hambourg, "Carleton Watkins," 10. On Watkins's big tree photographs, see also Elizabeth Hutchinson, "They Might Be Giants: Carleton Watkins, Galen Clark, and the Big Tree," in Alan C. Braddock and Christoph Irmscher, eds., *A Keener Perception: Ecocritical Studies in American Art History* (Tuscaloosa: University of Alabama Press, 2009), 110–28.

190. Quoted in Daniel J. Boorstin, *The Americans: The Democratic Experience* (New York: Random House, 1973), 170. On nineteenth-century Americans' passion for statistics, see "Statistical Communities," in Boorstin, *Americans*, 167–244; and Patricia Cline Cohen, "Statistics and the State: Changing Social Thought and the Emergence of a Quantitative Mentality in America, 1790–1820," *William and Mary Quarterly* 38, no. 1 (January 1981): 35–55.

191. On the Battle of the Wilderness, see McPherson, *Battle Cry of Freedom*, 724–28. On Americans' preoccupation with the statistics of the dead and wounded during the Civil War, see "Numbering: How Many? How Many?" in Faust, *This Republic of Suffering*, 250–65.

192. George Perkins Marsh, *Man and Nature: or, Physical Geography as Modified by Human Action* (New York: Charles Scribner, 1864), 43.

193. Quoted in Roderick Nash, *Wilderness and the American Mind*, 3rd ed. (New Haven, CT: Yale University Press, 1982), 105. On Marsh's ideas about forest preservation, see also Cox et al., *This Well-Wooded Land*, 144–47. On the wilderness in nineteenth-century American painting, see Angela L. Miller, "The Fate of Wilderness in American Landscape Art: The Dilemmas of 'Nature's Nation,' " in Braddock and Imscher, *Keener Perception*, 85–109. See also Huth, *Nature and the American*, 168–71.

194. Other marks on the tree in the painting include a heart, two sets of initials (BD or OD), and the number 17. For additional information on the painting, see Lucretia H. Geise, "Winslow Homer's Civil War Painting *The Initials*: A Little-Known Drawing and Related Works," *American Art Journal* 18, no. 3 (1986): 4–19. On women's use of the half-mourning dress during the Civil War, see Faust, *This Republic of Suffering*, 147–55.

195. At the end of the war, Lincoln put Barton in charge of correspondence about missing soldiers and identifying unmarked graves. On the role of female nurses in the Civil War, see McPherson, *Battle Cry of Freedom*, 479–84.

196. Lydia Howard Huntley Sigourney, *Scenes in My Native Land* (Boston: J. Munroe, 1845), 243.

197. Susan Fenimore Cooper, *Rural Hours by a Lady* (London: Putnam's American Agency, 1850), 128. On Cooper's ideas about the preservation of forests, see " 'Earth Is the Common Home of All': Susan Fenimore Cooper's Investigations of a Settled Landscape," in Michael A. Bryson, *Visions of the Land: Science, Literature, and the American Environment from the Era of Exploration to the Age of Ecology* (Charlottesville: University Press of Virginia, 2002), 105–33. The Civil War greatly accelerated the destruction of American forests, according to government reports of the period. The Commissioner of Agriculture noted that the "destruction of forests and timber during the war of rebellion has been immense" (*Report of the Commissioner of Agriculture for the Year 1865* [Washington, DC: Government Printing Office, 1866], 215). The Civil War also brought to center stage debates about the state of the nation's woodlands and conflicting ideas about their role in the United States.

198. Fitz Hugh Ludlow, *The Heart of the Continent* (New York: Hurd and Houghton; Cambridge, MA: Riverside Press, 1870), 424.

199. Quoted in Huth, *Nature and the American*, 141.

200. It is estimated that approximately three hundred thousand native people lived in the California region when Spanish settlers arrived in 1769, and the creation of the Yosemite preserve did not take into account the land's numerous native residents. The Yosemite area was the location of one of the most shameful episodes in nineteenth-century white-tribal relations. During the "Mariposa War," the U.S. Army's Mariposa Batallion attacked the Awahneechee people, a Yosemite group of the Miwok tribe. The soldiers killed unarmed men, women, and children, burned their villages, and expelled survivors from their ancestral lands. A member of the invading batallion explained cryptically after the battle that "[n]o prisoners were taken." Quoted in Robert H. Keller and Michael F. Turek, *American Indians and National Parks* (Tucson: University of Arizona Press, 1998), 21.

201. On Indian land management in California, see the excellent book by Thomas C. Blackburn and Kat Anderson, comp. and eds., *Before the Wilderness: Environmental Management by Native Californians* (Menlo Park, CA: Ballena Press, 1993), 18–19. On the larger subject of the relationship between Indian culture and the natural environment in California, see Robert F. Heizer and Albert B. Elsasser, *The Natural World of the California Indians* (Berkeley: University of California Press, 1980).

202. Quoted in Baker, *Heartless Immensity*, 2.

203. "The white man sure ruined this country," said James Rust, a Southern Sierra Miwok elder. "It's turned back to wilderness." California Indians believe that when people are gone from an area long enough, the continuity of knowledge, passed down through generations, is broken and the land becomes estranged from human concerns. See M. Kat Anderson, *Tending the Wild: Native American Knowledge and the Management of California's Natural Resources* (Berkeley: University of California Press, 2005), 3–4.

EPILOGUE

204. Cincinnati, where Duncanson made his home, was a haven for free blacks and fugitive slaves and the headquarters of numerous abolitionist organizations. The city nevertheless seethed with racial tensions and in 1841 was the site of the worst race riot in U.S. history to that date. Similar to other communities across the North, Cincinnati's trade associations accepted only white members, and free blacks were left with the lowest paying and most menial jobs. Reprisals against "overreaching" blacks who fought against the status quo were frequent events. On the subject of free blacks in the United States during the antebellum period, see Leon F. Litwak, *North of Slavery: The Negro in the Free States, 1790–1860* (Chicago: University of Chicago Press, 1961).

205. Church's rainbow was so realistic that the famous British art critic John Ruskin stepped in front of it, thinking it came from the refraction of sunlight on a nearby windowpane. On Ruskin's encounter with Church's rainbow and its international fame, see Jeremy Elwell Adamson, *Niagara: Two Centuries of Changing Attitudes, 1697–1901*, exh. cat. (Washington, DC: Corcoran Gallery of Art, 1985), 16.

206. Quoted in Joseph D. Ketner, *The Emergence of the African-American Artist: Robert S. Duncanson, 1821–1872* (Columbia: University of Missouri Press, 1993), 84, 88–89.

207. In his choice of a career as a fine artist, Duncanson faced formidable obstacles and hardships. Even his own family questioned his work, accusing the light-skinned artist of trying to pass as white. Duncanson vehemently denied the accusation, saying, "I have no color on the brain, all I have on the brain is paint." For additional information about Duncanson's struggles as a free black and a landscape artist, see Ketner, *Robert S. Duncanson*; "Reconstructing Duncanson," in David M. Lubin, *Picturing a Nation: Art and Social Change in Nineteenth-Century America* (New Haven, CT: Yale University Press, 1994), 107–57; and Guy McElroy, "Robert S. Duncanson: A Study of the Artist's Life and Work," in *Robert Duncanson: A Centennial Exhibition*, exh. cat. (Cincinnati: Cincinnati Art Museum, 1972), 1–17.

208. George Musser, "Rainbows: The Simple Magic of Their Shape and Colors Still Puzzles," *Scientific American* 301, no. 3 (September 2009): 70. On nineteenth-century American artists' ideas about painting rainbows, see "A Chapter on Rainbows in Landscapes," in *Crayon* 7, no. 2 (February 1860): 40–41. On the history of artists, writers, and scientists who study rainbows, see Raymond L. Lee Jr. and Alistair B. Fraser, *The Rainbow Bridge: Rainbows in Art, Myth, and Science* (University Park: Pennsylvania State University Press, 2001); and Richard Whelan, *The Book of Rainbows: Art, Literature, Science, and Mythology* (New York: First Glance Books, 1997).

✤ List of Illustrations

NIAGARA FALLS

FIG. 22
Arthur Lumley
Niagara Seen with Different
Eyes. From *Harper's Weekly*,
August 9, 1873

FIG. 23
John James Barralet
*Science Unveiling the Beauties of
Nature to the Genius of America*,
1814, wash drawing. Print
collection, Miriam and Ira D.
Wallach Division of Art, Prints
and Photographs, The New York
Public Library, Astor, Lenox and
Tilden Foundations

FIG. 24
Alexander Wilson
White-headed Eagle, about 1811,
proof plate. From *American
Ornithology; or The Natural
History of the United States*
(Philadelphia: Bradford and
Inskeep, 1808–14). The Academy
of Natural Sciences, Ewell Sale
Stewart Library and the Albert
M. Greenfield Digital Imaging
Center for Collections

FIG. 25
John James Barralet after
Alexander Wilson
View of the Great Pitch Taken
from Below. From *Port Folio*
3, no. 3 (March 1810). General
Research Division, The New York
Public Library, Astor, Lenox and
Tilden Foundations

FIG. 26
George Catlin
*Bird's Eye View of Niagara
Falls*, about 1827, gouache,
17 ⅝ × 15 ½ in. Private collection.
Photo courtesy of Hirschl &
Adler Galleries, New York

FIG. 27
John Trumbull
*Niagara Falls from Below the
Great Cascade on the British
Side*, 1808, oil on canvas,
24 ⁷⁄₁₆ × 36 ⅜ in. Wadsworth
Atheneum Museum of Art,
Hartford, Connecticut, Bequest
of Daniel Wadsworth, 1848.5.
Photo courtesy Wadsworth
Atheneum Museum of Art/Art
Resource, NY.

FIG. 28
Alvan Fisher
*The Great Horseshoe Fall,
Niagara*, 1820, oil on canvas,
34 ⅜ × 48 in. Smithsonian
American Art Museum,
Museum purchase, 1966.82.1

FIG. 29
Raphaelle Peale
*Venus Rising from the Sea—A
Deception*, about 1822, oil on
canvas, 29 ⅛ × 24 ⅛ in. The
Nelson-Atkins Museum of Art,
Kansas City, Missouri, Purchase:
William Rockhill Nelson Trust,
34-147. Photo by Jamison Miller

FIG. 30
Thomas Cole
*Horseshoe Falls, Niagara,
Morning*, about 1829, pen and
brown ink over graphite on off-
white wove paper, 9 ¹¹⁄₁₆ × 12 ³⁄₁₆ in.
The Detroit Institute of Arts,
USA, Founders Society Purchase,
William H. Murphy Fund, The
Bridgeman Art Library

FIG. 31
Thomas Cole
Distant View of Niagara Falls
1830, oil on panel, 18 ⅞ × 23 ⅞ in.
The Art Institute of Chicago,
Friends of American Art
Collection, 1946.396. Photo by
Robert Hashimoto. Photography
© The Art Institute of Chicago

FIG. 32
Samuel Finley Breese Morse
*Benjamin Silliman, B.A. 1796,
M.A. 1799*, 1825, oil on canvas,
55 ¼ × 44 ¼ in. Yale University
Art Gallery, Gift of Bartlett
Arkell, B.A. 1886, M.A. 1898,
to Silliman College

FIG. 33
Thomas Cole
*The Subsiding of the Waters of
the Deluge*, 1829, oil on canvas,
35 ¾ × 47 ¾ in. Smithsonian
American Art Museum, Gift of
Mrs. Katie Dean in memory of
Minnibel S. and James Wallace
Dean and museum purchase
through the Smithsonian
Institution Collections
Acquisition Program, 1983.40

FIG. 34
Frederic Edwin Church
*At the Base of the American
Falls, Niagara*, 1856, brush
and oil on paper laminate,
11 ⅝ × 13 ¾ in. Cooper Hewitt,
National Design Museum, Gift of
Louis P. Church, 1917-4-1353.
Photo by Matt Flynn. Photo
courtesy Cooper Hewitt,
National Design Museum/Art
Resource, NY

FIG. 35
John Rapkin
Niagara, United States, about 1851,
engraving. From Montgomery
Martin, *Illustrated Atlas, and
Modern History of the World*
(New York: John Tallis & Co.,
1851). 12 ⅝ × 9 ½ in. Castellani Art
Museum of Niagara University
Collection, Generous Donation
from Dr. Charles Rand Penney,
partially funded by the Castellani
Purchase Fund, with additional
funding from Mr. and Mrs.
Thomas A. Lytle, 2006

FIG. 36
Frederic Edwin Church
Niagara, 1857, oil on canvas,
42 ½ × 90 ½ in.
Corcoran Gallery of Art,
Washington, D.C., Museum
Purchase, Gallery Fund

FIG. 37
Thomas Doughty
*Mill Pond and Mills, Lowell,
Massachusetts*, about 1833, oil
on canvas, 26 × 35 in. Private
collection

FIG. 38
Unidentified artist
Why Doesn't the Water Drop
Off? From Jacob Abbott, *Rollo's
Philosophy: Water* (Boston:
Phillips, Sampson, 1855)

FIG. 39
Bass Otis
Interior of a Smithy, by 1815, oil on
canvas, 50 ⅝ × 80 ½ in. Courtesy of
the Pennsylvania Academy of the
Fine Arts, Philadelphia,
Gift of the artist

FIG. 40
Unidentified artist
Centre Vent Wheel at the Boott
Cotton Mills. From James B.
Francis, *Lowell Hydraulic
Experiments* (Boston: Little,
Brown, 1855)

FIG. 41
Cartoonist possibly Jacob Dallas
The Coming Man's Presidential
Career, à la Blondin. From
Harper's Weekly, August 25, 1860

FIG. 42
Unidentified artist
Patent application for Abraham
Lincoln's Device for Buoying
Vessels over Shoals, 1849.
National Archives and Records
Administration

THE BUFFALO

FIG. 68

Charles Willson Peale
Entrance Ticket to the Peale Museum, 1788, etching and engraving on cream wove paper, 2 11/16 × 3 3/4 in. Worcester Art Museum, Massachusetts, Richard A. Heald Fund and the Thomas Hovey Gage Fund

FIG. 69

George Catlin
Buffalo Bull, Grazing on the Prairie, 1832–33, oil on canvas, 24 × 29 in. Smithsonian American Art Museum, Gift of Mrs. Joseph Harrison, Jr.

FIG. 70

George Catlin
Buffalo Herds Crossing the Upper Missouri, 1832, oil on canvas, 11 1/4 × 14 3/8 in. Smithsonian American Art Museum, Gift of Mrs. Joseph Harrison, Jr.

FIG. 71

George Catlin.
Buffalo Hunt under the Wolf-skin Mask, 1832–33, oil on canvas, 24 × 29 in. Smithsonian American Art Museum, Gift of Mrs. Joseph Harrison, Jr.

FIG. 72

John Mix Stanley
Buffalo Hunt on the Southwestern Prairies, 1845, oil on canvas, 40 1/2 × 60 3/4 in. Smithsonian American Art Museum, Gift of the Misses Henry

FIG. 73

Felix Octavius Carr Darley
Buffalo Hunt. From John Frost, ed., *Book of the Indians of North America: Illustrating Their Manners, Customs, and Present State* (Hartford, CT: W. J. Hamersley, 1852)

FIG. 74

Unidentified artist
Copper Weather Vanes. From *Catalogue of the J. L. Mott Iron Works, New York and Chicago*, about 1892. Courtesy, The Winterthur Library, Printed Book and Periodical Collection

FIG. 75

After James Henry Moser
A Still Hunt, 1870s. From William T. Hornaday, *The Extermination of the American Bison* (Washington, D.C.: Government Printing Office, 1889)

FIG. 76

Ernest Griset
The Far West. —Shooting Buffalo on the Line of the Kansas-Pacific Railroad. From *Frank Leslie's Illustrated Newspaper*, June 3, 1871

FIG. 77

Kunkel Brothers, publishers. Cover for sheet music, "Kansas Pacific Grand March," 1872

FIG. 78

Thomas Nast
The Last Buffalo. From *Harper's Weekly*, June 6, 1874

FIG. 79

Albert Bierstadt
The Last of the Buffalo about 1888, oil on canvas, 71 1/8 × 118 3/4 in. Corcoran Gallery of Art, Washington, D.C., Gift of Mary (Mrs. Albert) Bierstadt

FIG. 80

Eadweard Muybridge
Buffalo: galloping, 1887, collotype, 8 3/4 × 13 7/8 in. Iris and B. Gerald Cantor Center for Visual Arts at Stanford University; Stanford Family Collections

FIG. 81

Henry Worrall
Taking and Being Taken. From William E. Webb, *Buffalo Land* (Cincinnati and Chicago: E. Hannaford, 1872). Private collection

A LOCOMOTIVE PEOPLE

FIG. 82

Thomas Worth
Currier and Ives, publishers. *A "Limited Express." "Five Seconds for Refreshments"!*, 1884, colored lithograph, 10 × 13 in. Museum of the City of New York, The Harry T. Peters Collection

FIG. 83

Unidentified artist after John B. Jervis, locomotive designer. Design for the locomotive Experiment which was renamed Brother Jonathan, 1831, drawing on paper. Transportation Collections, National Museum of American History, Smithsonian Institution, Washington, D.C.

FIG. 84

Designs for spark arrestors of the Baldwin Locomotive Works, 1860, lithograph

FIG. 85

Asher B. Durand
Railway Cut, n.d., engraving, 11 1/4 × 7 7/8 in. Print collection, Miriam and Ira D. Wallach Division of Art, Prints and Photographs, The New York Public Library, Astor, Lenox and Tilden Foundations

FIG. 86

Robert J. Havell Jr.
Two Artists in a Landscape about 1840–1850, oil on canvas, 35 × 50 in. New York State Historical Association, Cooperstown

FIG. 87

Unidentified artist
The Horrors of Travel. From *Harper's Weekly*, September 23, 1865

FIG. 88

Unidentified photographer
Artists' Excursion, Sir John's Run, Berkeley Springs, 1858, salt paper print, 6 3/4 × 6 1/8 in. Smithsonian American Art Museum, Museum purchase from the Charles Issacs Collection made possible in part by the Luisita L. and Franz H. Denghausen Endowment

FIG. 89

George Inness
The Lackawanna Valley about 1856, oil on canvas, 33 7/8 × 50 3/16 in. National Gallery of Art, Washington, D.C., Gift of Mrs. Huttleston Rogers. Image courtesy of the Board of Trustees

FIG. 90

Jasper Francis Cropsey
Starrucca Viaduct, Pennsylvania, 1865, oil on canvas, 22 3/8 × 36 3/8 in. Toledo Museum of Art, Purchased with funds from the Florence Scott Libbey Bequest in Memory of her Father, Maurice A. Scott, 1947.58

FIG. 91

Unidentified artist
Destruction by Rebels of the B&O Bridge over the Potomac River, Harpers Ferry, June 14, 1861. From *Harper's Weekly*, July 6, 1861

FIG. 92

Unidentified engraver after photograph by A. J. Russell. Dale Creek Bridge. From *Illustrated London News*, November 27, 1869

✦ Bibliography

Abbott, Jacob. *Rollo's Philosophy: Water*. Boston: Phillips, Sampson 1855.

Abir-Am, Pnina G., and Dorinda Outram, eds. *Uneasy Careers and Intimate Lives: Women in Science, 1789–1979*. New Brunswick, NJ: Rutgers University Press, 1987.

Adams, Bluford. *E Pluribus Barnum: The Great Showman and the Making of U.S. Popular Culture*. Minneapolis: University of Minnesota Press, 1997.

Adamson, Jeremy Elwell. *Niagara: Two Centuries of Changing Attitudes, 1697–1901*. Exh. cat. Washington, DC: Corcoran Gallery of Art, 1985.

Allen, Elsa Guerdrum. "The History of American Ornithology before Audubon." *Transactions of the American Philosophical Society* 41, pt. 3 (1951): 552–69.

Allen, J. A. *The American Bisons, Living and Extinct*. Cambridge, MA: University Press, 1876.

Allen, Thomas M. *A Republic in Time: Temporality and Social Imagination in Nineteenth-Century America*. Chapel Hill: University of North Carolina Press, 2008.

Allen, Zachariah. *The Science of Mechanics as Applied to the Present Improvements in the Useful Arts in Europe, and in the United States of America*. Providence, RI: Hutchens and Cory, 1829.

"American Miracle." *Poulson's American Daily Advertiser*, February 18, 1802.

Amram, Fred M. B. and Susan K. Henderson. *African-American Inventors*. Mankato, MN: Capstone Press, 1996.

Anderson, M. Kat. *Tending the Wild: Native American Knowledge and the Management of California's Natural Resources*. Berkeley: University of California Press, 2005.

Anderson, Nancy K., and Linda S. Ferber. *Albert Bierstadt: Art and Enterprise*. Exh. cat. New York: Hudson Hills Press in association with Brooklyn Museum, 1990.

Appel, Toby A. "Science, Popular Culture and Profit: Peale's Philadelphia Museum." *Journal of the Society for the Bibliography of Natural History* 9, no. 4 (April 1980): 619–34.

Appleby, Joyce. *Inheriting the Revolution: The First Generation of Americans*. Cambridge, MA: Belknap Press of Harvard University Press, 2001.

Armengaud, Jacques-Eugène, Charles Armengaud, and Jules Amouroux. *The Practical Draughtsman's Book of Industrial Design*. William Johnson trans. and ed. New York: Stringer and Townsend, 1857.

Arnold, Lois Barber. *Four Lives in Science: Women's Education in the Nineteenth Century*. New York: Schocken Books, 1984.

Asma, Stephen T. *Stuffed Animals and Pickled Heads: The Culture and Evolution of Natural History Museums*. New York: Oxford University Press, 2001.

Axelrod, Robert. *The Evolution of Cooperation*. New York: Basic Books, 1984.

Baigell, Matthew. *Albert Bierstadt*. New York: Watson-Guptill, 1981.

Bain, David Haward. *Empire Express: Building the First Transcontinental Railroad*. New York: Viking Books, 1999.

Baker, Anne. *Heartless Immensity: Literature, Culture, and Geography in Antebellum America*. Ann Arbor: University of Michigan Press, 2006.

Baldwin, Dwight, Jr., Judith De Luce, and Carl Pletsch, eds. *Beyond Preservation: Restoring and Inventing Landscapes*. Minneapolis: University of Minnesota Press, 1994.

Banta, Martha. *Taylored Lives: Narrative Productions in the Age of Taylor, Veblen, and Ford*. Chicago: University of Chicago Press, 1993.

Barber, Lynn. *The Heyday of Natural History, 1820–1870*. Garden City, NY: Doubleday, 1980.

Barnard, Henry. *Armsmear, The Home, the Arm, and the Armory of Samuel Colt: A Memorial*. 1866; rpt., Maryland: Beinfeld, 1976.

Barnum, Phineas Taylor (P. T.). *The Life of P. T. Barnum, Written by Himself*. 1855; rpt., Urbana: University of Illinois Press, 2000.

Baron, Ava, ed. *Work Engendered: Toward a New History of American Labor*. Ithaca, NY: Cornell University Press, 1991.

Barrow, Mark V., Jr. *Nature's Ghosts: Confronting Extinction from the Age of Jefferson to the Age of Ecology*. Chicago: University of Chicago Press, 2009.

Barsness, Larry. *The Bison in Art: A Graphic Chronicle of the American Bison*. Exh. cat. Flagstaff, AZ: Northland Press, 1977.

Bartky, Ian R. "The Adoption of Standard Time." *Technology and Culture* 30, no. 1 (January 1989): 25–56.

Battison, Edwin A. *Muskets to Mass Production: The Men and the Times That Shaped American Manufacturing*. Windsor, VT: American Precision Museum, 1976.

Beaver, Donald de B. "Altruism, Patriotism, and Science: Scientific Journals in the Early Republic." *American Studies* 12, no. 1 (1971): 5–19.

Becker, Lydia Ernestine. "On the Study of Science by Women." *Contemporary Review* 10 (January–April 1869): 386–404.

Bedell, Rebecca. "Thomas Cole and the Fashionable Science." *Huntington Library Quarterly: Studies in English and American History and Literature* 59, nos. 2 and 3 (1996): 348–78.

——. *The Anatomy of Nature: Geology and American Landscape Painting, 1825–1875*. Princeton, NJ: Princeton University Press, 2001.

Bedini, Silvio A. *Thinkers and Tinkers: Early American Men of Science*. New York: Charles Scribner, 1975.

——. *Thomas Jefferson: Statesman of Science*. New York: Macmillan, 1990.

Beecher, Catharine E. *A Treatise on Domestic Economy, for the Use of Young Ladies at Home, and at School*. Boston: Marsh, Capen, Lyon, and Webb, 1841.

Bell, Whitfield J., Jr. *A Cabinet of Curiosities: Five Episodes in the Evolution of American Museums*. Charlottesville: University Press of Virginia, 1967.

Bellesiles, Michael A. *Arming America: The Origins of a National Gun Culture*. New York: Alfred A. Knopf, 2000.

Bellesiles, Michael A., ed. *Lethal Imagination: Violence and Brutality in American History*. New York: New York University Press, 1999.

Benjamin, Park, ed. *Appletons' Cyclopædia of Applied Mechanics: A Dictionary of Mechanical Engineering and the Mechanical Arts*. 2 vols. New York: D. Appleton, 1880.

Bennis, Warren, and Patricia Ward Biederman. *Organizing Genius: The Secrets of Creative Collaboration*. New York: Basic Books, 1998.

Benson, Maxine. *Martha Maxwell, Rocky Mountain Naturalist*. Lincoln: University of Nebraska Press, 1986.

Benyus, Janine M. *Biomimicry: Innovation Inspired by Nature*. New York: Morrow, 1997.

Berns, Gregory. *Iconoclast: A Neuroscientist Reveals How to Think Differently*. Boston: Harvard Business School Press, 2008.

Berton, Pierre. *Niagara: A History of the Falls*. New York: Kodansha International, 1997.

Betts, John Rickards. "P. T. Barnum and the Populariztion of Natural History." *Journal of the History of Ideas* 20, no. 3 (June–September 1959): 353–68.

Bieder, Robert E. *Science Encounters the Indian, 1820–1880: The Early Years of American Ethnology*. Norman: University of Oklahoma Press, 1986.

Bigelow, Jacob. *Elements of Technology*. 2nd ed. Boston: Hilliard, Gray, Little, and Wilkins, 1831.

Bigelow, Timothy. *Journal of a Tour to Niagara Falls in the Year 1805*. Boston: Press of J. Wilson, 1876.

Bijker, Wiebe E., Thomas P. Hughes, and Trevor J. Pinch, eds. *The Social Construction of Technological Systems: New Directions in the Sociology and History of Technology*. Cambridge, MA: MIT Press, 1987.

Bishop, J. Leander. *A History of American Manufactures from 1608 to 1860…Comprising Annals of the Industry of the United States in Machinery, Manufactures and Useful Arts*. 2nd ed. 3 vols. Philadelphia: Edward Young, 1868.

Blackburn, Thomas C., and Kat Anderson, comp. and eds. *Before the Wilderness: Environmental Management by Native Californians*. Menlo Park, CA: Ballena Press, 1993.

Bleier, Ruth, ed. *Feminist Approaches to Science*. New York: Pergamon Press, 1986.

Blondheim, Menahem. *News over the Wires: The Telegraph and the Flow of Public Information in America, 1844–1897*. Cambridge, MA: Harvard University Press, 1994.

Blum, Ann Shelby. *Picturing Nature: American Nineteenth-Century Zoological Illustration*. Princeton, NJ: Princeton University Press, 1993.

Bode, Carl. *The American Lyceum: Town Meeting of the Mind*. New York: Oxford University Press, 1956.

Boehm, Christopher. *Hierarchy in the Forest: The Evolution of Egalitarian Behavior*. Cambridge, MA: Harvard University Press, 1999.

Boessenecker, John. *Gold Dust and Gunsmoke: Tales of Gold Rush Outlaws, Gunfighters, Lawmen, and Vigilantes*. New York: John Wiley, 1999.

Bolger, Doreen, Marc Simpson, and John Wilmerding, eds. *William M. Harnett*. Exh. cat. Fort Worth, TX: Amon Carter Museum of Western Art; New York: Metropolitan Museum of Art and Harry N. Abrams, 1992.

Boller, Paul F., Jr. *American Thought in Transition: The Impact of Evolutionary Naturalism, 1865–1900*. Edited by David D. Tassel. Chicago: Rand McNally, 1969.

Bonner, John Tyler. *Why Size Matters: From Bacteria to Blue Whales*. Princeton, NJ: Princeton University Press, 2006.

Bonta, Marcia Myers. *Women in the Field: America's Pioneering Women Naturalists*. College Station: Texas A&M University Press, 1991.

Booker, Margaret Moore. *Among the Stars: The Life of Maria Mitchell*. Nantucket, MA: Mill Hill Press, 2007.

Boorstin, Daniel J. *The Americans: The Democratic Experience*. New York: Random House, 1973.

———. *The Americans: The National Experience*. New York: Random House, 1965.

Borut, Michael. *The "Scientific American" in Nineteenth Century America*. Ann Arbor, MI: University Microfilms International, 1977.

Botkin, B. A., and Alvin F. Harlow, eds. *A Treasury of Railroad Folklore: The Stories, Tall Tales, Traditions, Ballads, and Songs of the American Railroad Man*. New York: Bonanza Books, 1953.

Bowker, Geoffrey C. *Memory Practices in the Sciences*. Cambridge, MA: MIT Press, 2005.

Bowles, Samuel. *Across the Continent: A Summer's Journey to the Rocky Mountains, the Mormons, and the Pacific States, with Speaker Colfax*. Springfield, MA: S. Bowles; New York: Hurd and Houghton, 1865; rpt., New York: Readex Microprint, 1968.

Braddock, Alan C., and Christoph Irmscher, eds. *A Keener Perception: Ecocritical Studies in American Art History*. Tuscaloosa: University of Alabama Press, 2009.

Branch, E. Douglas. *The Hunting of the Buffalo*. New York: D. Appleton, 1929.

Brigham, David R. "'Ask the Beasts, and They Shall Teach Thee': The Human Lessons of Charles Willson Peale's Natural History Display." *Huntington Library Quarterly: Studies in English and American History and Literature* 59, nos. 2 and 3 (1996): 182–206.

———. *Public Culture in the Early Republic: Peale's Museum and Its Audience*. Washington, DC: Smithsonian Institution Press, 1995.

Brockman, John. *The Third Culture: Beyond the Scientific Revolution*. New York: Touchstone, 1995.

Brockman, John, ed. *Curious Minds: How a Child Becomes a Scientist*. New York: Vintage Books, 2004.

Bronowski, Jacob. *Science and Human Values*. London: Faber and Faber, 1956.

Brown, Carolyn S. *The Tall Tale in American Folklore and Literature*. Knoxville: University of Tennessee Press, 1987.

Brown, Chandos Michael. *Benjamin Silliman: A Life in the Young Republic*. Princeton, NJ: Princeton University Press, 1989.

Brown, Richard D. *Modernization: The Transformation of American Life, 1600–1865*. American Century Series. New York: Hill and Wang, 1976.

Brown, Travis. *Historical First Patents: The First United States Patent for Many Everyday Things*. Metuchen, NJ: Scarecrow Press, 1994.

Bruce, Dickson D., Jr. *Violence and Culture in the Antebellum South*. Austin: University of Texas Press, 1979.

Bruce, Robert V. *The Launching of Modern American Science, 1846–1876*. New York: Knopf, 1987.

——. *Lincoln and the Tools of War*. Indianapolis, IN: Bobbs-Merrill, 1956.

Brückner, Martin. *The Geographic Revolution in Early America: Maps, Literacy, and National Identity*. Chapel Hill: Published for Omohundro Institute of Early American History and Culture by University of North Carolina Press, 2006.

Bryson, Michael A. *Visions of the Land: Science, Literature, and the American Environment from the Era of Exploration to the Age of Ecology*. Charlottesville: University Press of Virginia, 2002.

Buell, Lawrence. *The Environmental Imagination: Thoreau, Nature Writing, and the Formation of American Culture*. Cambridge, MA: Belknap Press of Harvard University Press, 1995.

Burlingame, Roger. *Inventors Behind the Inventor*. New York: Harcourt, Brace and World, 1947.

——. *Machines That Built America*. New York: Signet Books, 1953.

Butler, David. F. *United States Firearms: The First Century, 1776–1875*. New York: Winchester Press, 1971.

Byrne, Oliver. *The American Engineer, Draftsman, and Machinist's Assistant: Designed for Practical Workingmen, Apprentices, and Those Intended for the Engineering Profession*. Philadelphia: C. A. Brown, 1853.

Cannon, Susan Faye. *Science in Culture: The Early Victorian Period*. Kent, UK: Dawson; New York: Science History, 1978.

Carter, Gregg Lee, ed. *Guns in American Society: An Encyclopedia of History, Politics, Culture, and the Law*. 2 vols. Santa Barbara, CA: ABC-CLIO, 2002.

Catlin, George. *Letters and Notes on the Manners, Customs, and Conditions of North American Indians: Written during Eight Years' Travel (1832–1839) amongst the Wildest Tribes of Indians in North America*. 2 vols. 1841; rpt., New York: Dover, 1973.

Chambers, Bruce W. *Old Money: American Trompe L'Oeil Images of Currency*. Exh. cat. New York: Berry-Hill Galleries, 1988.

Chandler, Alfred D., Jr. *The Visible Hand: The Managerial Revolution in American Business*. Cambridge, MA: Belknap Press of Harvard University Press, 1977.

Chandler, Alfred D., Jr., and James W. Cortada, eds. *A Nation Transformed by Information: How Information Has Shaped the United States from Colonial Times to the Present*. New York: Oxford University Press, 2000.

Chant, Christopher. *The History of North American Steam*. Edison, NJ: Chartwell Books, 2007.

Chaplin, Joyce E. "Review: The Curious Case of Science and Empire." *Reviews in American History* 34, no. 4 (December 2006): 434–40.

Chevalier, Michael. *Society, Manners, and Politics in the United States; Being a Series of Letters on North America*. Trans. from 3rd Paris ed. Boston: Weeks, Jordan, 1839.

Chevalier, Michel. *Society, Manners, and Politics in the United States; Letters on North America*. Edited by John W. Ward. Translated by T. G. Bradford. Gloucester, MA: Peter Smith, 1967.

Cikovsky, Nicolai, Jr. *George Inness*. New York: Harry N. Abrams; Washington, DC: National Museum of American Art, Smithsonian Institution, 1993.

——. "George Inness and the Hudson River School: *The Lackawanna Valley*," *American Art Journal* 2, no. 2 (Fall 1970): 36–57.

——. " 'The Ravages of the Axe': The Meaning of the Tree Stump in Nineteenth-Century American Art." *Art Bulletin* 61, no. 4 (December 1979): 611–26.

——. "Winslow Homer's (So-called) *Morning Bell*." *American Art Journal* 29, nos. 1 and 2 (1998): 4–17.

Cincinnati Art Museum. *Robert S. Duncanson: A Centennial Exhibition*. Cincinnati: Cincinnati Art Museum, 1972.

Clark, Charles. *Brainstorming: How to Create Sucessful Ideas*. North Hollywood, CA: Melvin Powers Wilshire Book, 1989.

Clark, Christopher. *Social Change in America: From the Revolution through the Civil War*. Chicago: Ivan R. Dee, 2006.

Clark, Galen. *Indians of the Yosemite Valley and Vicinity: Their History, Customs, and Traditions*. 1904; rpt., Charleston, SC: Forgotten Books, 2008.

Coen, Rena N. "The Last of the Buffalo." *American Art Journal* 5, no. 2 (November 1973): 833–94.

Cohen, I. Bernard. *Science and the Founding Fathers: Science in the Political Thought of Jefferson, Franklin, Adams, and Madison*. New York: W. W. Norton, 1995.

——. *Some Early Tools of American Science: An Account of the Early Scientific Instruments and Mineralogical and Biological Collections in Harvard University*. Cambridge: Harvard University Press, 1950.

Cohen, Patricia Cline. "Statstics and the State: Changing Social Thought and the Emergence of a Quantitative Mentality in America, 1790–1820." *William and Mary Quarterly* 38, no. 1 (January 1981): 35–55.

Combs, Barry B. *Westward to Promontory: Building the Union Pacific across the Plains and Mountains: A Pictorial Documentary.* New York: Crown, 1986.

Cook, James W. *The Arts of Deception: Playing with Fraud in the Age of Barnum.* Cambridge, MA: Harvard University Press, 2001.

Cooke, Josiah Parsons. *Scientific Culture and Other Essays.* 2nd ed. New York: D. Appleton, 1885.

Cooper, Carolyn C. *Shaping Invention: Thomas Blanchard's Machinery and Patent Management in Nineteenth-Century America.* New York: Columbia University Press, 1991.

Cooper, James Fenimore. *The Pioneers, or, The Sources of the Susquehanna: A Descriptive Tale.* 1823; rpt., New York: Signet Classics, 2007.

Cooper, Jeff. *The Pistol as a Weapon of Defence in the House and on the Road: How to Choose It and How to Use It.* 1875; rpt., New York: Paladin Press, 2004.

Cooper, Susan Fenimore. *Rural Hours by a Lady.* 1850; rpt. New York: General Books, 2009.

Cooper-Hewitt Museum. *American Enterprise: Nineteenth-Century Patent Models.* New York: Cooper-Hewitt Museum, Smithsonian Institution, 1984.

Cornell, Saul. *A Well-Regulated Militia: The Founding Fathers and the Origins of Gun Control in America.* New York: Oxford University Press, 2006.

Corner, James, and Alex S. MacLean. *Taking Measures: Across the American Landscape.* New Haven, CT: Yale University Press, 1996.

Cott, Nancy F. *The Bonds of Womanhood: "Woman's Sphere" in New England, 1780–1835.* New Haven, CT: Yale University Press, 1977.

Coultas, Harland. *What May Be Learned from a Tree.* New York: D. Appleton, 1860.

Cowan, Ruth Schwartz. *A Social History of American Technology.* New York: Oxford University Press, 1997.

Cox, Thomas R., Joseph J. Malone, Roberts S. Maxwell, and Phillip Drennon Thomas. *This Well-Wooded Land: Americans and Their Forests from Colonial Times to the Present.* Lincoln: University of Nebraska Press, 1985.

Craven, Wayne. "Asher B. Durand's Career as an Engraver." *American Art Journal* 3, no. 1 (Spring 1971): 39–57.

Cravens, Hamilton, Alan I. Marcus, and David M. Katzman, eds. *Technical Knowledge in American Culture: Science, Technology, and Medicine Since the Early 1800s.* Tuscaloosa: University of Alabama Press, 1996.

Cronon, William. *Changes in the Land: Indians, Colonists, and the Ecology of New England.* New York: Hill and Wang, 1983.

——. *Nature's Metropolis: Chicago and the Great West.* New York: W. W. Norton, 1991.

——, ed. *Uncommon Ground: Rethinking the Human Place in Nature.* New York: W. W. Norton, 1996.

Csikszentmihalyi, Mihaly. *Creativity: Flow and the Psychology of Discovery and Invention.* New York: Harper Collins, 1996.

Custer, General George Armstrong. *My Life on the Plains, or, Personal Experiences with Indians.* Norman: University of Oklahoma Press, 1962.

Daily, Gretchen C., and Katherine Ellison. *The New Economy of Nature: The Quest to Make Conservation Profitable.* Washington, DC: Island Press/Shearwater Books, 2002.

Daniels, George H. *American Science in the Age of Jackson.* Tuscaloosa: University of Alabama Press, 1994.

Daniels, George H., ed. *Darwinism Comes to America.* Waltham, MA: Blaisdell, 1968.

——. *Nineteenth-Century American Science: A Reappraisal.* Evanston, IL: Northwestern University Press, 1972.

Danly, Susan, and Leo Marx, eds. *The Railroad in American Art: Representations of Technological Change.* Cambridge, MA: MIT Press, 1988.

Dary, David A. *The Buffalo Book: The Full Saga of the American Animal.* Athens: Swallow Press/Ohio University Press, 1989.

Dasgupta, Subrata. *Technology and Creativity*. New York: Oxford University Press, 1996.

Daston, Lorraine, and Katharine Park. *Wonders and the Order of Nature, 1150-1750*. New York: Zone Books, 1998.

Daston, Lorraine, and Peter Galison. "The Image of Objectivity." *Representations* 40 (Fall 1992): 81-128.

Davidson, Abraham. "Catastrophism and Peale's 'Mammoth.' " *American Quarterly* 21 (1969): 620-29.

Deakin, Roger. *Wildwood: A Journey Through Trees*. New York: Free Press, 2008.

Dear, Peter. *The Intelligibility of Nature: How Science Makes Sense of the World*. Chicago: University of Chicago Press, 2006.

Delbourgo, James. *A Most Amazing Scene of Wonders: Electricity and Enlightenment in Early America*. Cambridge, MA: Harvard University Press, 2006.

Dennis, Mary B. *A Study of Leaves*. New York: D. Appleton, 1888.

DiChristina, Mariette, et al. "Let Your Creativity Soar." *Scientific American Mind* 19, no. 3 (June/July 2008): 24-31.

Dizard, Jan E., Robert Merrill Muth, and Stephen P. Andrews Jr., eds. *Guns in America: A Reader*. New York: New York University Press, 1999.

Dobyns, Kenneth W. *The Patent Office Pony: A History of the Early Patent Office*. Fredericksburg, VA: Sergeant Kirkland's Museum and Historical Society, 1994.

Donald, David Herbert, ed. *Gone for a Soldier: The Civil War Memoirs of Private Alfred Bellard*. Boston: Little, Brown, 1975.

Dood, Kendall J. "Patent Models and the Patent Law: 1790-1880." Pt. 2, Conclusion. *Journal of the Patent Office Society* 65, no. 5 (May 1983): 232-85.

Douglas, George H. *All Aboard! The Railroad in American Life*. New York: Marlowe, 1995.

Drucker, Johanna. "Harnett, Haberle, and Peto: Visuality and Artifice among the Proto-Modern Americans." *Art Bulletin* (March 1992): 37-50.

Dublin, Thomas. *Transforming Women's Work: New England Lives in the Industrial Revolution*. Ithaca, NY: Cornell Universty Press, 1994.

Dugatkin, Lee Alan. *Mr. Jefferson and the Giant Moose: Natural History in Early America*. Chicago: University of Chicago Press, 2009.

Dunaway, Finis. *Natural Visions: The Power of Images in American Environmental Reform*. Chicago: University of Chicago Press, 2005.

Dunlap, William. *History of the Rise and Progress of the Arts of Design in the United States*. 3 vols. 1834; rev. and enlarged by Frank W. Bayley and Charles Goodspeed. New York: B. Blom, 1965.

Dunlop, M. H. "Curiosities Too Numerous to Mention: Early Regionalism and Cincinnati's Western Museum." *American Quarterly* 36, no. 4 (Fall 1984): 524-48.

Dupree, A. Hunter, ed. *Science and the Emergence of Modern America, 1865-1916*. Berkeley Series in American History. Edited by Charles Sellers. Chicago: Rand McNally, 1963.

Durand, John. *The Life and Times of Asher B. Durand*. 1894; rpt., Hensonville, NY: Black Dome, 2006.

Dutton, Denis. *The Art Instinct: Beauty, Pleasure, and Human Evolution*. New York: Bloomsbury Press, 2009.

Eager, Gerald. "The Image of the Bison: Changing Perceptions of the American West." *Nineteenth Century* 14, no. 2 (1994): 26-37.

Ede, Siân. *Art and Science*. London: I. B. Tauris, 2005.

Edwards, David. *Artscience: Creativity in the Post-Google Generation*. Cambridge, MA: Harvard University Press, 2008.

Eisler, Benita, ed. *The Lowell Offering: Writings by New England Mill Women (1840-1845)*. Philadelphia: Lippincott, 1977.

Ellis, Joseph J. *After the Revolution: Profiles of Early American Culture*. New York: W. W. Norton, 1979.

Engbeck, Joseph H., Jr. *The Enduring Giants: The Epic Story of Giant Sequoia and the Big Trees of Calaveras*. Berkeley: University Extension, University of California, Berkeley, 1973.

Evans, Harold, et al. *They Made America: From the Steam Engine to the Search Engine, Two Centuries of Innovators*. New York: Little, Brown, 2004.

Faust, Drew Gilpin. *This Republic of Suffering: Death and the American Civil War*. New York: Alfred A. Knopf, 2008.

Feist, Gregory J. *The Psychology of Science and the Origins of the Scientific Mind*. New Haven, CT: Yale University Press, 2006.

"Female Inventive Talent." *Scientific American* 23, no. 12 (September 17, 1870): 184.

Fenster, Julie M. *The Spirit of Invention: The Story of the Thinkers, Creators, and Dreamers Who Formed Our Nation*. Washington, DC: Smithsonian Books, 2009.

Ferber, Linda S., ed. *Kindred Spirits: Asher B. Durand and the American Landscape*. New York: Brooklyn Museum, 2007.

Ferguson, Eugene S. "The American-ness of American Technology." *Technology and Culture* 20, no. 1 (January 1979): 3–24.

——. "The Mind's Eye: Nonverbal Thought in Technology." *Science* 197, no. 4306 (August 26, 1977): 827–36.

Fisher, Marvin. "The Iconology of Industrialism, 1830–60," *American Quarterly* 13, no. 3 (Fall 1961): 347–64.

Fisher, Marvin. *Workshops in the Wilderness: The European Response to American Industrialization, 1830–1860*. New York: Oxford University Press, 1967.

Fletcher, Malcolm, and John Taylor, eds. *Railways: The Pioneer Years*. Secaucus, NJ: Chartwell Books, 1990.

Flores, Dan. "Bison Ecology and Bison Diplomacy: The Southern Plains from 1800 to 1850." *Journal of American History* 78, no. 2 (September 1991): 465–85.

Florida, Richard. *The Rise of the Creative Class and How It's Transforming Work, Leisure, Community, and Everyday Life*. New York: Basic Books, 2004.

Fogel, Robert William. *The Fourth Great Awakening and the Future of Egalitarianism*. Chicago: University of Chicago Press, 2000.

Foley, Paul J. *Willard's Patent Time Pieces: A History of the Weight-Driven Banjo Clock, 1800–1900*. Norwell, MA: Roxbury Village, 2002.

Folsom, Michael Brewster, and Stephen D. Lubar, eds. *The Philosophy of Manufactures: Early Debates over Industrialization in the United States*. Cambridge, MA: MIT Press, 1982.

Foner, Philip S., ed. *The Factory Girls: A Collection of Writings on Life and Struggles in the New England Factories of the 1840's by the Factory Girls Themselves, and the Story, in Their Own words, of the First Trade Unions of Women Workers in the United States*. Urbana: University of Illinois Press, 1977.

Ford, Charlotte A. "Eliza Frances Andrews, Practical Botanist, 1840–1931." *Georgia Historical Quarterly* 70, no. 1 (Spring 1986): 63–80.

Forester, Frank. *The Complete Manual for Young Sportsmen*. New York: Stringer and Townsend, 1857.

Foss, Phillip O., ed. *Conservation in the United States: A Documentary History; Recreation*. New York: Chelsea House, 1971.

Foster, David R., and John D. Aber, eds. *Forests in Time: The Environmental Consequences of One Thousand Years of Change in New England*. New Haven, CT: Yale University Press, 2004.

Francis, James B. *Lowell Hydraulic Experiments, Being a Selection from Experiments on Hydraulic Motors, on the Flow of Water over Weirs, and in Canals of Uniform Rectangular Section and of Short Length*. Boston: Little, Brown, 1855.

Franke, Mary Ann. *To Save the Wild Bison: Life on the Edge in Yellowstone*. Norman: University of Oklahoma Press, 2005.

Freeman, Joanne B. "Dueling as Politics: Reinterpreting the Burr-Hamilton Duel." *William and Mary Quarterly* 53, no. 2 (April 1996): 289–318.

Fyfe, Aileen, and Bernard Lightman, eds. *Science in the Marketplace: Nineteenth-Century Sites and Experiences*. Chicago: University of Chicago Press, 2007.

Gage, Matilda Joslyn. "Woman as Inventor." *North American Review* 136, no. 318 (May 1883): 478–89.

Garavaglia, Louis A., and Charles G. Worman. *Firearms of the American West, 1803–1865*. Niwot: University Press of Colorado, 1998.

Gardner, Howard. *Multiple Intelligences: The Theory in Practice*. New York: Basic Books, 1993.

Geise, Lucretia H. "Winslow Homer's Civil War Painting *The Initials*: A Little-Known Drawing and Related Works," *American Art Journal* 18, no. 3 (1986): 4–19.

Geist, Valerius. *Buffalo Nation: History and Legend of the North American Bison*. Minnesota: Voyageur Press, 1996.

Geological Survey of California. *The Yosemite Guide-book: A Description of the Yosemite Valley and the Adjacent Region of the Sierra Nevada, and of the Big Trees of California*. Cambridge, MA: University Press, Welch, Bigelow, 1870.

George-Warren, Holly, and Michelle Freedman. *How the West Was Worn*. New York: Harry N. Abrams in association with Autry Museum of Western Heritage, Los Angeles, 2001.

Getzels, Jacob W., and Mihaly Csikszentmihalyi. "Scientific Creativity." *Science Journal* 3, no. 9 (September 1967): 80–84.

——. *The Creative Vision: A Longitudinal Study of Problem Finding in Art*. New York: John Wiley, 1976.

Gibbons, Felton, and Deborah Strom. *Neighbors to the Birds: A History of Birdwatching in America*. New York: W. W. Norton, 1988.

Giedion, Siegfried. *Mechanization Takes Command: A Contribution to Anonymous History*. New York: Oxford University Press, 1948.

Gies, Joseph, and Frances Gies. *The Ingenious Yankees: The Men, Ideas, and Machines That Transformed a Nation, 1776–1876*. New York: Thomas Y. Crowell, 1976.

Gillispie, Charles Coulston. *Genesis and Geology: A Study in the Relations of Scientific Thought, Natural Theology, and Social Opinion in Great Britain, 1790–1850*. Cambridge, MA: Harvard University Press, 1996.

Godman, John Davidson. *American Natural History*. Vol. 1, pt. 1, Mastology. 1826–28; rpt., New York: Arno Press, 1974.

Goetzmann, William H. *Army Exploration in the American West, 1803–1863*. New Haven, CT: Yale University Press, 1959.

——. *Exploration and Empire: The Explorer and the Scientist in the Winning of the American West*. New York: W. W. Norton, 1966.

Gordon, Robert B. "Custom and Consequence: Early Nineteenth-Century Origins of the Environmental and Social Costs of Mining Anthracite." In *Early American Technology: Making and Doing Things from the Colonial Era to 1850*, edited by Judith A. McGaw, 240–77. Chapel Hill: University of North Carolina Press for Institute of Early American History and Culture, Williamsburg, Virginia, 1994.

Greeley, Horace. *An Overland Journey from New York to San Francisco in the Summer of 1859*. San Francisco: H. H. Bancroft, 1860.

Greeley, Horace, et al. *The Great Industries of the United States: Being an Historical Summary of the Origin, Growth, and Perfection of the Chief Industrial Arts of This Country*. Hartford, CT: J. B. Burr and Hyde, 1872.

Green, Constance McLaughlin. *Eli Whitney and the Birth of American Technology*. Edited by Oscar Handlin. New York: Longman, 1956.

Greenblatt, Stephen. *Marvelous Possessions: The Wonder of the New World*. Chicago: University of Chicago Press, 1991.

——. "Resonance and Wonder." In Ivan Karp and Steven D. Lavine, eds., *Exhibiting Cultures: The Poetics and Politics of Museum Display*, 42–56. Washington, DC: Smithsonian Institution Press, 1991.

Greene, John C. "American Science Comes of Age, 1780–1820." *Journal of American History* 55, no. 1 (June 1968): 22–41.

——. *American Science in the Age of Jefferson*. Ames: Iowa State University Press, 1984.

Greene, Mott T. *Geology in the Nineteenth Century: Changing Views of a Changing World*. Ithaca, NY: Cornell University Press, 1982.

Gross, Ernie. *Advances and Innovations in American Daily Life, 1600–1930s*. Jefferson, NC: McFarland, 2002.

Gurney, George, and Therese Thau Heyman, eds. *George Catlin and His Indian Gallery*. Washington, DC: Smithsonian American Art Museum; New York: W. W. Norton, 2002.

Haber, Louis. *Black Pioneers of Science and Invention*. New York: Harcourt, Brace, and World, 1970.

Hagen, Joel B. "Amateurs, Professionals, and the Growth of American Botany." *Reviews in American History* 22 (1994): 133–37.

Hales, Peter B. *William Henry Jackson and the Transformation of the American Landscape*. Philadelphia: Temple University Press, 1988.

Hall, Dawn, ed. *Drawing the Borderline: Artist-Explorers of the U. S.-Mexico Boundary Survey*. Exh. cat. Albuquerque, NM: Albuquerque Museum of Art and History, 1996.

Hamblyn, Richard. *The Invention of Clouds: How an Amateur Meteorologist Forged the Language of the Skies*. New York: Farrar, Straus, and Giroux, 2001.

Hareven, Tamara K. *Family Time and Industrial Time: The Relationship between the Family and Work in a New England Industrial Community*. Cambridge: Cambridge University Press, 1982.

Harris, Neil. *Humbug: The Art of P. T. Barnum*. Boston: Little, Brown, 1973.

——. *The Artist in American Society: The Formative Years, 1790–1860*. Chicago: University of Chicago Press, 1982.

Harter, Jim. *American Railroads of the Nineteenth Century: A Pictoral History in Victorian Wood Engravings*. Lubbock, TX: Texas Tech University Press, 1998.

Hartigan, Lynda Roscoe. *Sharing Traditions: Five Black Artists in Nineteenth-Century America, from the Collections of the National Museum of American Art*. Washington, DC: Smithsonian Institution Press, 1985.

Harvey, Eleanor Jones. *The Painted Sketch: American Impressions from Nature, 1830–1880*. Exh. cat. Dallas, TX: Dallas Museum of Art in association with Harry N. Abrams, 1998.

Hasselstrom, Linda M. *Bison: Monarch of the Plains*. Portland, OR: Graphic Arts Center, 1998.

Hatch, Alden. *Remington Arms in American History*. New York: Rinehart, 1956.

Haven, Charles T., and Frank A. Belden. *A History of the Colt Revolver and the Other Arms Made by Colt's Patent Fire Arms Manufacturing Company from 1836 to 1940*. New York: Morrow, 1940.

Hawke, David Freeman. *Nuts and Bolts of the Past: A History of American Technology, 1776–1860*. New York: Harper and Row, 1988.

Hawken, Paul, Amory Lovins, and L. Hunter Lovins. *Natural Capitalism: Creating the Next Industrial Revolution*. New York: Little, Brown, 1999.

Hazen, Margaret Hindle, and Robert M. Hazen. *Wealth Inexhaustible: A History of America's Mineral Industries to 1850*. New York: Van Nostrand Reinhold, 1985.

Hazen, Robert M., and Margaret Hindle Hazen. *American Geological Literature, 1669 to 1850*. Stroudsburg, PA: Dowden, Hutchinson, and Ross, 1980.

Hedeen, Stanley. *Big Bone Lick: The Cradle of American Paleontology*. Lexington: University Press of Kentucky, 2008.

Heilbroner, Robert L. "Do Machines Make History?" *Technology and Culture* 8, no. 3 (July 1967): 335–45.

Hendricks, Gordon. *Albert Bierstadt: Painter of the American West*. New York: Harrison House, 1988.

Henson, Pamela M. "Objects of Curious Research: The History of Science and Technology at the Smithsonian." *Isis* 90, Supplement (June 1999): 249–69.

Herman, Daniel Justin. *Hunting and the American Imagination*. Washington, DC: Smithsonian Institution Press, 2001.

Hewes, Jeremy Joan. *Redwoods: The World's Largest Trees*. Chicago: Rand McNally, 1981.

Hindle, Brooke. *Emulation and Invention*. New York: New York University Press, 1981.

——. "The Exhilaration of Early American Technology: An Essay." In *Early American Technology: Making and Doing Things from the Colonial Era to 1850*, edited by Judith A. McGaw, 40-67. Chapel Hill: University of North Carolina Press for Institute of Early American History and Culture, Williamsburg, Virginia, 1994.

——. *The Pursuit of Science in Revolutionary America, 1735–1789*. Chapel Hill: University of North Carolina Press for Institute of Early American History and Culture, Williamsburg, Virginia, 1956.

Hindle, Brooke, ed. *Material Culture of the Wooden Age*. Tarrytown, NY: Sleepy Hollow Press, 1981.

Hindle, Brooke, and Steven Lubar. *Engines of Change: The American Industrial Revolution, 1790–1860*. Washington, DC: Smithsonian Institution Press, 1986.

Hoffer, Eric. *The True Believer: Thoughts on the Nature of Mass Movements*. New York: Harper Perennial, 1951.

Holyoak, Keith J., and Paul Thagard. *Mental Leaps: Analogy in Creative Thought*. Cambridge, MA: MIT Press, 1995.

Hornaday, William Temple. *The Extermination of the American Bison*, 1889. Rpt. Washington, DC: Smithsonian Institution Press, 2002.

Hosley, William. *Colt: The Making of an American Legend*. Amherst: University of Massachusetts Press, 1996.

Hounshell, David A. *From the American System to Mass Production, 1800–1932: The Development of Manufacturing Technology in the United States*. Baltimore: Johns Hopkins University Press, 1984.

Houze, Herbert G. *Samuel Colt: Arms, Art, and Invention*. Edited by Elizabeth Mankin Kornhauser. New Haven, CT: Yale University Press and Wadsworth Atheneum Museum of Art, 2006.

Hovenkamp, Herbert. *Science and Religion in America, 1800–1860*. Philadelphia: University of Pennsylvania Press, 1978.

Hughes, James B., comp. *The Gatling Gun Notebook: A Collection of Data and Illustrations*. Lincoln, RI: Andrew Mowbray, 2000.

Hughes, Thomas Parke. *Human-Built World: How to Think about Technology and Culture*. Chicago: University of Chicago Press, 2004.

Hughes, Thomas Parke, ed. *Changing Attitudes toward American Technology*. New York: Harper and Row, 1975.

Hunter, Louis C. *A History of Industrial Power in the United States, 1780–1930*. Vol. 1, *Waterpower in the Century of the Steam Engine*. Charlottesville: University Press of Virginia for Eleutherian Mills–Hagley Foundation, 1979.

——. *A History of Industrial Power in the United States, 1780–1930*. Vol. 2, *Steam Power*. Charlottesville: University Press of Virginia Eleutherian Mills-Hagley Foundation, 1985.

Hurtado, Albert L. *Indian Survival on the California Frontier*. New Haven, CT: Yale University Press, 1988.

Huth, Hans. *Nature and the American: Three Centuries of Changing Attitudes*. 2nd ed. Lincoln: University of Nebraska Press, 1991.

Impey, Oliver, and Arthur MacGregor, eds., *The Origins of Museums: The Cabinet of Curiosities in Sixteenth- and Seventeenth-Century Europe*. Oxford: Clarendon Press; New York: Oxford University Press, 1985.

Ingersoll, Charles J. *A Discourse Concerning the Influence of America on the Mind; Being the Annual Oration Delivered before the American Philosophical Society, at the University in Philadelphia, on the 18th October, 1823*. Philadelphia: A. Small, 1823.

Isenberg, Andrew C. *The Destruction of the Bison: An Environmental History, 1750–1920*. Cambridge: Cambridge University Press, 2000.

——. "Toward a Policy of Destruction: Buffaloes, Law, and the Market, 1803–83." *Great Plains Quarterly* 12, no. 4 (Fall 1992): 227–41.

Jackson, John N. *The Mighty Niagara: One River—Two Frontiers*. Amherst, NY: Prometheus Books, 2003.

Jacobus, Mary, Evelyn Fox Keller, and Sally Shuttleworth, eds. *Body/Politics: Women and the Discourses of Science*. New York: Routledge, 1990.

James, Portia P. *The Real McCoy: African-American Invention and Innovation, 1619–1930*. Exh. cat. Washington, DC: Smithsonian Institution Press for Anacostia Museum, Smithsonian Institution, 1989.

Janssen, Barbara Suit, ed. *Icons of Invention: American Patent Models*. Washington, DC: National Museum of American History, Smithsonian Institution, 1990.

Jardine, N., J. A. Secord, and E. C. Spary, eds. *Cultures of Natural History*. Cambridge: Cambridge University Press, 1996.

Jarves, James Jackson. *Art Thoughts: The Experiences and Observations of an American Amateur in Europe*. New York: Hurd and Houghton, 1870.

Jefferson, Thomas. *Notes on the State of Virginia*, 1787; rpt., New York: Penguin Classics, 1988.

Jenkins, Reese V. *Images and Enterprise: Technology and the American Photographic Industry, 1839 to 1925*. Baltimore: Johns Hopkins University Press, 1975.

Jerome, Chauncey. *History of the American Clock Business for the Past Sixty Years, and Life of Chauncey Jerome, Written by Himself: Barnum's Connection with the Yankee Clock Business.* New Haven, CT: F. C. Dayton Jr., 1860.

Johnson, Steven. *Where Good Ideas Come From: The Natural History of Innovation.* New York: Riverhead Books, 2010.

Johns, Elizabeth. *American Genre Painting: The Politics of Everyday Life.* New Haven, CT: Yale University Press, 1991.

Jordan, Philip D. *Frontier Law and Order: Ten Essays.* Lincoln: University of Nebraska Press, 1970.

Judson, Lewis Van Hagen. *Weights and Measures Standards of the United States: A Brief History.* National Bureau of Standards Special Publication. Washington, DC: U.S. Government Printing Office, 1963.

Kanigel, Robert. *The One Best Way: Frederick Winslow Taylor and the Enigma of Efficiency.* New York: Viking, 1997.

Kasson, John F. *Civilizing the Machine: Technology and Republican Values in America, 1776–1900.* New York: Grossman, 1976.

Kates, Don B., Jr., ed. *Firearms and Violence: Issues of Public Policy.* San Francisco: Pacific Institute for Public Policy, 1997.

Kates, Don B., Jr., and Gary Kleck. *The Great American Gun Debate: Essays on Firearms and Violence.* San Francisco, CA: Pacific Research Institute for Public Policy Research, 1984.

Kao, John. *Innovation Nation: How America Is Losing Its Innovative Edge, Why It Matters, and What We Can Do to Get It Back.* New York: Free Press, 2007.

Keeney, Elizabeth B. *The Botanizers: Amateur Scientists in Nineteenth-Century America.* Chapel Hill: University of North Carolina Press, 1992.

Keller, Evelyn Fox. *A Feeling for the Organism: The Life and Work of Barbara McClintock.* San Francisco: W. H. Freeman, 1983.

——. *Making Sense of Life: Explaining Biological Development with Models, Metaphors, and Machines.* Cambridge, MA: Harvard University Press, 2002.

——. *Reflections on Gender and Science.* New Haven, CT: Yale University Press, 1985.

Keller, Julia. *Mr. Gatling's Terrible Marvel: The Gun That Changed Everything and the Misunderstood Genius Who Invented It.* New York: Viking, 2008.

Keller, Robert H., and Michael F. Turek. *American Indians and National Parks.* Tucson: University of Arizona Press, 1998

Kelley, Tom. *The Ten Faces of Innovation: IDEO's Strategies for Beating the Devil's Advocate and Driving Creativity throughout Your Organization.* New York: Currency/Doubleday, 2005.

Kelly, Franklin. *Frederic Edwin Church.* Exh. cat. Washington, DC: National Gallery of Art and Smithsonian Institution Press, 1989.

Kelly, Franklin, and Gerald L. Carr. *The Early Landscapes of Frederic Edwin Church, 1845–1854.* Fort Worth, TX: Amon Carter Museum of Western Art, 1987.

Kelsey, Darwin P., ed. *Farming in the New Nation: Interpreting American Agriculture, 1790–1840.* Washington, DC: Agricultural History Society, 1972.

Kemp, Martin. *Seen/Unseen: Art, Science, and Intuition from Leonardo to the Hubble Telescope.* New York: Oxford University Press, 2006.

——. *The Science of Art: Optical Themes in Western Art from Brunelleschi to Seurat.* New Haven, CT: Yale University Press, 1990.

Kennedy, Ian. "Crossing Continents: American and Beyond." In Ian Kennedy and Julian Treuherz, eds. *The Railway: Art in the Age of Steam.* Exh. cat. New Haven, CT: Yale University Press; Kansas City, MO: Nelson-Atkins Museum of Art; Liverpool: Walker Art Gallery, National Museums Liverpool, 2008.

Kennett, Lee, and James La Verne Anderson. *The Gun in America: The Origins of a National Dilemma.* Westport, CT: Greenwood Press, 1975.

Kenseth, Joy, ed. *The Age of the Marvelous.* Exh. cat. Hanover, NH: Hood Museum of Art, Dartmouth College, 1991.

Kessler-Harris, Alice. *Out to Work: A History of Wage-Earning Women in the United States.* New York: Oxford University Press, 2003.

Ketner, Joseph D. *The Emergence of the African-American Artist: Robert S. Duncanson, 1821–1872.* Columbia: University of Missouri Press, 1993.

Ketner, Joseph D., II. "Robert S. Duncanson (1821–1872): The Late Literary Landscape Paintings." *American Art Journal* 15, no. 1 (Winter 1983): 35–47.

Khan, B. Zorina. *The Democratization of Invention: Patents and Copyrights in American Economic Development, 1790–1920*. Cambridge: Cambridge University Press, 2005.

Klein, Maury. *The Power Makers: Steam, Electricity, and the Men Who Invented Modern America*. New York: Bloomsbury Press, 2008.

Klingender, Francis D. *Art and the Industrial Revolution*. Edited and revised by Arthur Elton. London: Evelyn, Adams, and Mackay, 1968.

Kohlstedt, Sally Gregory. "Curiosities and Cabinets: Natural History Museums and Education on the Antebellum Campus." *Isis* 79, no. 3 (September 1988): 405–26.

——. *The Formation of the American Scientific Community: The American Association for the Advancement of Science, 1848–60*. Urbana: University of Illinois Press, 1976.

——. "Maria Mitchell and the Advancement of Women in Science." In Pnina G. Abir-Am and Dorinda Outram, eds. *Uneasy Careers and Intimate Lives: Women in Science, 1789–1979*. New Brunswick, NJ: Rutgers University Press, 1987.

——. "Parlors, Primers, and Public Schooling: Education for Science in Nineteenth-Century America." *Isis* 81, no. 3 (September 1990): 424–45.

——. "In from the Periphery: American Women in Science, 1830–1880." Special Issue: Women, Science, and Society. *Signs: Journal of Women in Culture and Society* 4, no. 1 (August 1978): 81–96.

Kohlstedt, Sally Gregory, ed. *The Origins of Natural Science in America: The Essays of George Brown Goode*. Washington, DC: Smithsonian Institution Press, 1991.

Korzenik, Diana. *Drawn to Art: A Nineteenth-Century American Dream*. Hanover, NH: University Press of New England, 1985.

Krensky, Stephen. *What's the Big Idea? Four Centuries of Innovation in Boston*. Watertown, MA: Charlesbridge and Boston History Collaborative, 2008.

Kruska, Dennis G. *Sierra Nevada Big Trees: History of the Exhibitions, 1850–1903*. Los Angeles: Dawson's Book Shop, 1985.

Kulik, Gary. "Designing the Past: History Museum Exhibitions from Peale to the Present." In *History Museums in the United States: A Critical Assessment*, edited by Warren Leon and Roy Rosenzweig. Urbana: University of Illinois Press, 1989.

Kuritz, Hyman. "The Popularization of Science in Nineteenth-Century America." *History of Education Quarterly* 21, no. 3 (Fall 1981): 259–74.

Lane, Christopher W. *Impressions of Niagara: The Charles Rand Penney Collection of Prints of Niagara Falls and the Niagara River from the Sixteenth to the Early Twentieth Century*. Philadelphia: Philadelphia Print Shop, 1993.

Langdon, William Chauncy. *Everyday Things in American Life, 1607–1776*. 2 vols. New York: Charles Scribner, 1937.

Larkin, Jack. *The Reshaping of Everyday Life, 1790–1840*. New York: Harper and Row, 1988.

Larkin, Oliver W. *Samuel F. B. Morse and American Democratic Art*. Boston: Little, Brown, 1954.

Latour, Bruno, and Steve Woolgar. *Laboratory Life: The Construction of Scientific Facts*. Princeton, NJ: Princeton University Press, 1986.

Layton, Edwin T., Jr., ed. *Technology and Social Change in America*. New York: Harper and Row, 1973.

Lears, T. J. Jackson. *No Place of Grace: Antimodernism and the Transformation of American Culture, 1880–1920*. Chicago: University of Chicago Press, 1994.

Lee, Raymond L., Jr., and Alistair B. Fraser. *The Rainbow Bridge: Rainbows in Art, Myth, and Science*. University Park: Pennsylvania State University Press, 2001.

Lemmon, J. G. "A Botanical Wedding Trip." *Californian* 4, no. 24 (December 1881): 517–25.

Lepore, Jill. "Not So Fast. Scientific Management Started as a Way to Work. How Did It Become a Way of Life?" *New Yorker* (October 12, 2009): 84–122.

——. "Our Own Devices: Does Technology Drive History?" *New Yorker* (May 12, 2008): 84–122.

Lerman, Nina E., Ruth Oldenziel, and Arwen P. Mohun, eds. *Gender and Technology: A Reader*. Baltimore: Johns Hopkins University Press, 2003.

Lerner, Gerda. "The Lady and the Mill Girl: Changes in the Status of Women in the Age of Jackson." *American Studies Journal* 10 (April 1969): 5–15.

Levy, Lester S. *Picture the Songs: Lithographs from the Sheet Music of Nineteenth-Century America*. Baltimore: Johns Hopkins University Press, 1976.

Lewis, Andrew J. "A Democracy of Facts, An Empire of Reason: Swallow Submersion and Natural History in the Early American Republic." *William and Mary Quarterly*, 3rd ser., 62, no. 4 (October 2005): 663–96.

Lewis, M.D., Elisha J. *The American Sportsman: Containing Hints to Sportsmen, Notes on Shooting*. Philadelphia: J. B. Lippincott, 1863.

Lewis, Oscar. *The Big Four: The Story of Huntington, Stanford, Hopkins, and Crocker, and of the Building of the Central Pacific Railroad*. New York: A. A. Knopf, 1938.

Lewis, Robert M., ed. *From Traveling Show to Vaudeville: Theatrical Spectacle in America, 1830–1910*. Baltimore: Johns Hopkins University Press, 2003.

Licht, Walter. *Working for the Railroad: The Organization of Work in the Nineteenth Century*. Princeton, NJ: Princeton University Press, 1983.

Lilly, Lambert. *The American Forest; or, Uncle Philip's Conversations with the Children about the Trees of America*. New York: Harper and Brothers, 1834.

Limerick, Patricia Nelson. *The Legacy of Conquest: The Unbroken Past of the American West*. New York: W. W. Norton, 1987.

Linklater, Andro. *Measuring America: How an Untamed Wilderness Shaped the United States and Fulfilled the Promise of Democracy*. New York: Walker, 2002.

Lipman, Jean. *Rufus Porter Rediscovered: Artist, Inventor, Journalist, 1792–1884*. New York: Clarkson N. Potter, 1980.

Litwak, Leon F. *North of Slavery: The Negro in the Free States, 1790–1860*. Chicago: University of Chicago Press, 1961.

Livingstone, Margaret. *Vison and Art: The Biology of Seeing*. New York: Harry N. Abrams, 2002.

Loewenberg, Bert James. "The Controversy over Evolution in New England, 1859–1873." *New England Quarterly* 8, no. 2 (June 1935): 232–57.

Long, Stephen H. *Rail Road Manual, or a Brief Exposition of Principles and Deductions Applicable in Tracing the Route of a Rail Road*. Pt. 1. Baltimore: W. M. Wooddy, 1829.

Lott, Dale F. *American Bison: A Natural History*. Berkeley: University of California Press, 2002.

Lovejoy, Arthur O. *The Great Chain of Being: A Study of the History of an Idea*. The William James Lectures delivered at Harvard University, 1933. Cambridge, MA: Harvard University Press, 1936.

Lubin, David M. *Picturing a Nation: Art and Social Change in Nineteenth-Century America*. New Haven, CT: Yale University Press, 1994.

Ludlow, Fitz Hugh. *The Heart of the Continent: A Record of Travel across the Plains and in Oregon, with an Examination of the Mormon Principle*. New York: Hurd and Houghton; Cambridge, MA: Riverside Press, 1870.

Lurie, Edward. "Science in American Thought." *Journal of World History* 8 (1965): 638–65.

Lyden, Anne M. *Railroad Vision: Photography, Travel, and Perception*. Exh. cat. Los Angeles: J. Paul Getty Museum, 2003.

Mabee, Carleton. *The American Leonardo: A Life of Samuel F. B. Morse*. Rev. ed. Fleischmanns, NY: Purple Mountain Press, 2000.

Macdonald, Anne L. *Feminine Ingenuity: Women and Invention in America*. New York: Ballantine Books, 1992.

MacKenzie, Donald, and Judy Wajcman, eds. *The Social Shaping of Technology*. 2nd ed. Philadelphia: Open University Press, 1999.

Mann, Maybelle. "Francis William Edmonds: Mammon and Art." *American Art Journal* 2, no. 2 (Fall 1970): 92–106.

Marcus, Alan I., and Howard P. Segal. *Technology in America: A Brief History*. San Diego: Harcourt Brace Jovanovich, 1989.

Markham, George. *Guns of the Wild West—Firearms of the American Frontier, 1849–1917: The Handguns, Longarms, and Shotguns of the Gold Rush, the American Civil War, the Wild West, and the Armed Forces*. New York: Arms and Armour Press, 1991.

Markman, Arthur B., and Kristin L. Wood. *Tools for Innovation: The Science Behind the Practical Methods That Drive New Ideas*. New York: Oxford University Press, 2009.

Marsh, George Perkins. *Man and Nature; or, Physical Geography as Modified by Human Action*. New York: Charles Scribner, 1864.

Marvin, Carolyn. *When Old Technologies Were New: Thinking about Electric Communication in the Late Nineteenth Century*. New York: Oxford University Press, 1988.

Marzio, Peter C. *The Art Crusade: An Analysis of American Drawing Manuals, 1820–1860*. Washington, DC: Smithsonian Institution Press, 1976.

Marx, Leo. *The Machine in the Garden: Technology and the Pastoral Ideal in America*. New York: Oxford University Press, 1964.

——. *The Pilot and the Passenger: Essays on Literature, Technology, and Culture in the United States*. New York: Oxford University Press, 1988.

Mayr, Otto, and Robert C. Post, eds. *Yankee Enterprise: The Rise of the American Systems of Manufactures*. Washington, DC: Smithsonian Institution Press, 1981.

McGaw, Judith A., ed. *Early American Technology: Making and Doing Things from the Colonial Era to 1850*. Chapel Hill: University of North Carolina Press for Institute of Early American History and Culture, Williamsburg, Virginia, 1994.

McGraw, Donald J. "The Tree That Crossed a Continent." *California History* 61, no. 2 (Summer 1982): 120–39.

McKelvey, Susan Delano. *Botanical Exploration of the Trans-Mississippi West, 1790–1850*. Jamaica Plain, MA: Arnold Arboretum of Harvard University, 1955.

McKibben, Bill. *Eaarth: Making a Life on a Tough New Planet*. New York: Time Books, 2011.

——. *The End of Nature*. New York: Random House, 1989.

McKinsey, Elizabeth. *Niagara Falls: Icon of the American Sublime*. Cambridge: Cambridge University Press, 1985.

McMurry, Linda O. *George Washington Carver, Scientist and Symbol*. New York: Oxford University Press, 1981.

McPherson, James M. *Battle Cry of Freedom: The Civil War Era*. New York: Oxford University Press, 1988.

Meier, Hugo A. "Technology and Democracy, 1800–1860." *Mississippi Valley Historical Review* 43, no. 4 (March 1957): 618–40.

Merchant, Carolyn. *The Columbia Guide to American Environmental History*. New York: Columbia University Press, 2002.

——. *The Death of Nature: Women, Ecology, and the Scientific Revolution*. New York: Harper and Row, 1980.

——. *Ecological Revolutions: Nature, Gender, and Science in New England*. Chapel Hill: University of North Carolina Press, 1989.

Merrill, George P. *The First One Hundred Years of American Geology*. New York: Hafner, 1964.

Merritt, Raymond H. *Engineering in American Society, 1850–1875*. Lexington: University Press of Kentucky, 1969.

Miller, Char, ed. *American Forests: Nature, Culture, and Politics*. Lawrence: University Press of Kansas, 1997.

Miller, Lillian B. "Charles Willson Peale as History Painter: *The Exhumation of the Mastodon*." *American Art Journal* 13, no. 1 (Winter 1981): 47–68.

——. *The Lazzaroni: Science and Scientists in Mid-Nineteenth-Century America*. Exh. cat. Washington, DC: Smithsonian Institution Press for National Portrait Gallery, 1972.

Miller, Lillian B., ed. *The Peale Family: Creation of a Legacy, 1770–1870*. New York: Abbeville Press in association with Trust for Museum Exhibitions and National Portrait Gallery, Smithsonian Institution, 1996.

Miller, Lillian B., and David C. Ward, eds. *New Perspectives on Charles Willson Peale: A 250th Anniversary Celebration*. Pittsburgh: University of Pittsburgh Press for Smithsonian Institution, 1991.

Miller, Ralph N. "American Nationalism as a Theory of Nature." *William and Mary Quarterly* 12, no. 1 (January 1955): 74–95.

Mills, Paul Chadbourne. *The Buffalo Hunter and Other Related Versions of the Subject in Nineteenth-Century American Art and Literature*. Santa Barbara, CA: Santa Barbara Museum of Art, 2005.

Milner, Clyde A., II, Carol A. O'Connor, and Martha A. Sandweiss, eds. *The Oxford History of the American West*. New York: Oxford University Press, 1994.

Mitchell, Lee Clark. *Witnesses to a Vanishing America: The Nineteenth-Century Response*. Princeton, NJ: Princeton University Press, 1981.

Montgomery, Scott L. *Minds for the Making: The Role of Science in American Education, 1750–1990*. New York: Guilford Press, 1994.

Moreno, Jonathan D., and Rick Weiss, eds. *Science Next: Innovation for the Common Good from the Center for American Progress*. New York: Bellevue Literary Press, 2009.

Morison, Elting E. *From Know-How to Nowhere: The Development of American Technology*. New York: Basic Books, 1974.

Mumford, Lewis. *Technics and Civilization*. New York: Harcourt, Brace, 1934.

Murray, Daniel. "Who Invented the Cotton Gin?" *Voice of the Negro* 2, no. 2 (February 1905): 96–102.

Musser, George. "Rainbows: The Simple Magic of Their Shape and Colors Still Puzzles." *Scientific American* 301, no. 3 (September 2009): 70.

Nadis, Fred. *Wonder Shows: Performing Science, Magic, and Religion in America*. New Brunswick, NJ: Rutgers University Press, 2005.

Naef, Weston J. *Era of Exploration: The Rise of Landscape Photography in the American West, 1860–1885*. Buffalo: Albright-Knox Art Gallery, 1975.

National Parks and the American Landscape. Washington, DC: Smithsonian Institution Press for National Collection of Fine Arts, 1972.

Nelson, Daniel. *Frederick W. Taylor and the Rise of Scientific Management*. Madison: University of Wisconsin Press, 1980.

Nickel, Douglas R. *Carleton Watkins: The Art of Perception*. Exh. cat. San Francisco: San Francisco Museum of Modern Art; New York: Harry N. Abrams, 1999.

Noble, David F. *America by Design: Science, Technology, and the Rise of Corporate Capitalism*. New York: Knopf, 1977.

Novak, Barbara. "The Double-Edged Axe." *Art in America* 64 (January–February 1976): 44–50.

Nye, David E. *America as Second Creation: Technology and Narratives of New Beginnings*. Cambridge, MA: MIT Press, 2003.

——. *American Technological Sublime*. Cambridge, MA: MIT Press, 1994.

——. *Consuming Power: A Social History of American Energies*. Cambridge, MA: MIT Press, 1998.

——. *Electrifying America: Social Meanings of a New Technology, 1880–1940*. Cambridge, MA: MIT Press, 1990.

——. *Narratives and Spaces: Technology and the Construction of American Culture*. New York: Columbia University Press, 1997.

Nye, David E., ed. *Technologies of Landscape: From Reaping to Recycling*. Amherst: University of Massachusetts Press, 1999.

Nye, Russel B. *Society and Culture in America, 1830–1860*. New York: Harper and Row, 1974.

Nygren, Edward J. "The Almighty Dollar: Money as a Theme in American Painting." *Winterthur Portfolio* 23, nos. 2 and 3 (Summer/Autumn 1988): 129–50.

——*Views and Visions: American Landscape before 1830*. Exh. cat. Washington, DC: Corcoran Gallery of Art, 1986.

Oldenziel, Ruth. *Making Technology Masculine: Men, Women, and Modern Machines in America, 1870–1945*. Amsterdam, Netherlands: Amsterdam University Press, 1999.

Oleson, Alexandra, and Sanborn C. Brown, eds. *The Pursuit of Knowledge in the Early American Republic: American Scientific and Learned Societies from Colonial Times to the Civil War*. Baltimore: Johns Hopkins University Press, 1976.

O'Malley, Michael. *Keeping Watch: A History of American Time*. New York: Viking, 1990.

Ostrom, John H. "Mr. Peale's Missing Mastodon: The Story of an Ice Age Mastodon and Early American Painter Charles Willson Peale." *Discovery* 17, no. 1 (1983–84): 3–9.

Otter, Floyd L. *The Men of Mammoth Forest: A Hundred-Year History of a Sequoia Forest and Its People in Tulare County, California*. Ann Arbor, MI: Edwards Brothers, 1963.

Owen, Roberta J. M. "George Harvey's Anglo-American Atmospheric Landscapes." *Magazine Antiques* 174, no. 4 (October 2009): 112–21.

Pauly, Philip J. *Biologists and the Promise of American Life: From Meriwether Lewis to Alfred Kinsey*. Princeton, NJ: Princeton University Press, 2000.

———. *Fruits and Plains: The Horticultural Transformation of America*. Cambridge, MA: Harvard University Press, 2007.

Perry, Claire. "The Great American Experiment," *American Art* 23, no. 2 (Summer 2009): 16–20.

Pessen, Edward. *Jacksonian America: Society, Personality, and Politics*. Rev. ed. Urbana: University of Illinois Press, 1978.

Phelps, Mrs. [Almira H. L.]. *Botany for Beginners: An Introduction to Mrs. Lincoln's Lectures on Botany*. Stereotype edition. New York: F. J. Huntington, and Mason Brothers, 1856.

Phipps, Barry. *Beyond Measure: Conversations across Art and Science*. Exh. cat. Cambridge: University of Cambridge for Kettle's Yard, 2008.

Pink, Daniel H. *A Whole New Mind: Moving from the Information Age to the Conceptual Age*. New York: Riverhead Books, 2005.

Porter, Charlotte M. *The Eagle's Nest: Natural History and American Ideas, 1812–1842*. Tuscaloosa: University of Alabama Press, 1986.

Prime, Samuel Irenaeus. *The Life of Samuel F. B. Morse, LL.D., Inventor of the Electro-Magnetic Recording Telegraph*. New York: D. Appleton, 1875.

Prince, Sue Ann, ed. *Stuffing Birds, Pressing Plants, Shaping Knowledge: Natural History in North America, 1730–1860*. Philadelphia: American Philosophical Society, 2003.

Prodger, Phillip. *Time Stands Still: Muybridge and the Instantaneous Photography Movement*. Exh. cat. New York: Oxford University Press for Cantor Center for Visual Arts, Stanford University, 2003.

Pursell, Carroll W. *The Machine in America: A Social History of Technology*. Baltimore: Johns Hopkins University Press, 1995.

———. "Women Inventors in America." *Technology and Culture* 22, no. 3 (July 1981): 545–49.

Pursell, Carroll W., Jr., ed. *Technology in America: A History of Individuals and Ideas*. 2nd ed. Cambridge, MA: MIT Press, 1990.

Rae, John B. "The 'Know-How' Tradition: Technology in American History." *Technology and Culture* 1, no. 2 (Spring 1960): 139–50.

Ray, William, and Marlys Ray. *The Art of Invention: Patent Models and Their Makers*. Princeton, NJ: Pyne Press, 1974.

Reese, David Meredith. *Humbugs of New York: Being a Remonstrance Against Popular Delusion; Whether in Science, Philosophy, or Religion*. New York: John S. Taylor, 1838.

Regis, Pamela. *Describing Early America: Bartram, Jefferson, Crèvecoeur, and the Rhetoric of Natural History*. DeKalb: Northern Illinois University Press, 1992.

Reingold, Nathan. "Science in the Civil War: The Permanent Commission of the Navy Department." *Isis* 49, no. 3 (September 1958): 307–18.

———. *Science, American Style*. New Brunswick: Rutgers University Press, 1991.

Reingold, Nathan, ed. *Science in Nineteenth-Century America: A Documentary History*. New York: Hill and Wang, 1964.

———. *The Sciences in the American Context: New Perspectives*. Washington, DC: Smithsonian Institution Press, 1979.

Rennie, James. *Alphabet of Botany, for the Use of Beginners*. New York: Mahlon Day, 1837.

"Repeating Fire Arms." *United States Magazine* 4, no. 3 (March 1857): 221–49.

Revie, Linda L. *The Niagara Companion: Explorers, Artists, and Writers at the Falls, from Discovery through the Twentieth Century*. Waterloo, ON: Wilfrid Laurier University Press, 2003.

Rhodehamel, John. *The Great Experiment: George Washington and the American Republic*. Exh. cat. New Haven, CT: Yale University Press; San Marino, CA: Huntington Library, 1998.

Richardson, Edgar P., Brooke Hindle, and Lillian B. Miller. *Charles Willson Peale and His World*. New York: Harry N. Abrams, 1983.

Ridgway, Robert. *A Nomenclature of Colors for Naturalists*. Boston: Little, Brown, 1885.

Rigal, Laura. "Empire of Birds: Alexander Wilson's American Ornithology." *Huntington Library Quarterly: Studies in English and American History and Literature* 59, nos. 2 and 3 (1996): 232–68.

——. *The American Manufactory: Art, Labor, and the World of Things in the Early Republic*. Princeton, NJ: Princeton University Press, 1998.

Rinella, Steven. *American Buffalo: In Search of a Lost Icon*. New York: Spiegel and Grau, 2009.

Robertson, Charles J. *Temple of Invention: History of a National Landmark*. Exh. cat. Washington, DC: Smithsonian American Art Museum and National Portrait Gallery in association with Scala, 2006.

Robertson, David. *Yosemite as We Saw It: A Centennial Collection of Early Writings and Art*. Yosemite National Park, CA: Yosemite Association, 1990.

Rodgers, Daniel T. *The Work Ethic in Industrial America, 1850–1920*. Chicago: University of Chicago Press, 1978.

Rogow, Arnold A. *A Fatal Friendship: Alexander Hamilton and Aaron Burr*. New York: Hill and Wang, 1998.

Root-Bernstein, Robert, and Michèle Root-Bernstein. *Sparks of Genius: The Thirteen Thinking Tools of the World's Most Creative People*. Boston: Houghton Mifflin, 1999.

Rosen, Jonathan. *The Life of the Skies: Birding at the End of Nature*. New York: Farrar, Straus and Giroux, 2008.

Rosenberg, Nathan. *Technology and American Economic Growth*. New York: Harper and Row, 1972.

Rossiter, Margaret W. "Benjamin Silliman and the Lowell Institute: The Popularization of Science in Nineteenth-Century America." *New England Quarterly* 44, no. 4 (December 1971): 602–26.

——. *The Emergence of Agricultural Science: Justus Liebig and the Americans, 1840–1880*. New Haven, CT: Yale University Press, 1975.

——. *Women Scientists in America: Struggles and Strategies to 1940*. Baltimore: Johns Hopkins University Press, 1982.

——. " 'Women's Work' in Science, 1880–1910." *Isis* 71, no. 3 (September 1980): 381–98.

Royster, Charles. *A Revolutionary People at War: The Continental Army and American Character, 1775–1783*. Chapel Hill: University of North Carolina Press for Institute of Early American History and Culture, Williamsburg, Virginia, 1979.

Rudolph, Emanuel D. "Almira Hart Lincoln Phelps (1793–1884) and the Spread of Botany in Nineteenth Century America." *American Journal of Botany* 71, no. 8 (September 1984): 1161–67.

——. "Women in Nineteenth Century American Botany: A Generally Unrecognized Constituency." *American Journal of Botany* 69, no. 8 (September 1982): 1346–55.

Rudwick, Martin J. S. "The Emergence of a Visual Language for Geological Science, 1760–1840." *History of Science* 14, no. 25, prt. 3 (September 1976).

——. *Scenes from Deep Time: Early Pictorial Representations of the Prehistoric World*. Chicago: University of Chicago Press, 1992.

Russett, Cynthia Eagle. *Darwin in America: The Intellectual Response, 1865–1912*. San Francisco: W. H. Freeman, 1976.

Sacco, Ellen. "Racial Theory, Museum Practice: The Colored World of Charles Willson Peale." *Museum Anthropology* 20, no. 2 (September 1996): 25–32.

Sachs, Aaron. *The Humboldt Current: Nineteenth-Century Exploration and the Roots of American Environmentalism*. New York: Viking, 2006.

Sale, Kirkpatrick. *The Fire of His Genius: Robert Fulton and the American Dream*. New York: Free Press, 2001.

Sandoz, Mari. *The Buffalo Hunters: The Story of the Hide Men*. New York: Hastings House, 1954.

Sargent, C[harles] S[prague]. *The Woods of the United States*. New York: D. Appleton, 1885.

Schiebinger, Londa. *Nature's Body: Gender in the Making of Modern Science*. Brunswick, NJ: Rutgers University Press, 2004.

——. "The Philsopher's Beard: Women and Gender in Science." In *Cambridge History of Science. Volume 4: The Eighteenth Century*. Edited by Roy Porter. New York and Cambridge: Cambridge University Press, 2003.

Schivelbusch, Wolfgang. *The Railway Journey: The Industrialization of Time and Space in the Nineteenth Century*. Berkeley: University of California Press, 1986.

Schulten, Susan. *The Geographical Imagination in America, 1880-1950*. Chicago: University of Chicago Press, 2001.

Scott, Donald M. "The Popular Lecture and the Creation of a Public in Mid-Nineteenth-Century America." *Journal of American History* 66, no. 4 (March 1980): 791-809.

Scranton, Philip. *Endless Novelty: Specialty Production and American Industrialization, 1865-1925*. Princeton, NJ: Princeton University Press, 1997.

Sears, John F. *Sacred Places: American Tourist Attractions in the Nineteenth Century*. New York: Oxford University Press, 1989.

Seely, Bruce E. "Technology and Early American History." *Reviews in American History* 23, no. 4 (December 1995): 593-599.

Segal, Howard P. *Technological Utopianism in American Culture*. Twentieth Anniversary Edition. Syracuse, NY: Syracuse University Press, 2005.

Sellers, Charles Coleman. *Mr. Peale's Museum: Charles Willson Peale and the First Popular Museum of Natural Science and Art*. New York: W. W. Norton, 1980.

Sellers, Charles Grier. *The Market Revolution: Jacksonian America, 1815-1846*. New York: Oxford University Press, 1991.

Semonin, Paul. *American Monster: How the Nation's First Prehistoric Creature Became a Symbol of National Identity*. New York: New York University Press, 2000.

Sennett, Richard. *The Craftsman*. New Haven, CT: Yale University Press, 2008.

Serven, James E. *Two Hundred Years of American Firearms*. Chicago: Follett, 1975.

Silliman, Benjamin, and C. R. Goodrich, eds. *The World of Science, Art and Industry Illustrated from Examples in the New-York Exhibition, 1853-54*. New York: G. P. Putnam, 1854.

Silverman, Kenneth. *Lightning Man: The Accursed Life of Samuel F. B. Morse*. New York: Alfred A. Knopf, 2003.

Simonton, Dean Keith. *Origins of Genius: Darwinian Perspectives on Creativity*. New York: Oxford University Press, 1999.

Simpson, Marc, et al. *Winslow Homer: Paintings of the Civil War*. Exh. cat. San Francisco: Bedford Arts for Fine Arts Museums of San Francisco, 1988.

Sinclair, Bruce, ed. *Technology and the African-American Experience: Needs and Opportunities for Study*. Cambridge, MA: MIT Press, 2004.

Siracusa, Carl. *A Mechanical People: Perceptions of the Industrial Order in Massachusetts, 1815-1880*. Middletown, CT: Wesleyan University Press, 1979.

Slotkin, Richard. *Regeneration through Violence: The Mythology of the American Frontier, 1600-1860*. Middletown, CT: Wesleyan University Press, 1973.

——. *The Fatal Environment: The Myth of the Frontier in the Age of Industrialization, 1800-1890*. Middletown, CT: Wesleyan University Press, 1986.

Sluby, Patricia Carter. *The Inventive Spirit of African Americans: Patented Ingenuity*. Westport, CT: Praeger, 2004.

Smallwood, William Martin. *Natural History and the American Mind*. New York: Columbia University Press Press, 1941.

Smith, Cyril Stanley. *From Art to Science: Seventy-Two Objects Illustrating the Nature of Discovery*. Cambridge, MA: MIT Press, 1980.

——. "On Arts, Invention, and Technology." *Technology Review* 78, no. 7 (June 1976): 36-42.

——. "Art, Technology, and Science: Notes on Their Historical Interaction." *Technology and Culture* 11, no. 4 (October 1970): 493-549.

Smith, Jonathan. *Charles Darwin and Victorian Visual Culture*. Cambridge: Cambridge University Press, 2006.

Smith, Mark M. *Listening to Nineteenth-Century America*. Chapel Hill: University of North Carolina Press, 2001.

——. *Mastered by the Clock: Time, Slavery, and Freedom in the American South*. Chapel Hill: University of North Carolina Press, 1997.

Society for the History of Technology. Special Issue: Patents and Invention. *Technology and Culture* 32, no. 4 (October 1991).

Solnit, Rebecca. *River of Shadows: Eadweard Muybridge and the Technological Wild West*. New York: Viking, 2003.

Sokoloff, Kenneth. "Inventive Activity in Early Industrial America: Evidence from Patent Records, 1790-1846," *Journal of Economic History* 48, no. 4 (December 1988): 813-50.

"Special Report: Sustainability. Managing Earth's Future: Solutions for a Finite World." *Scientific American* 302, no. 4 (April 2010).

Spence, Mark David. *Dispossessing the Wilderness: Indian Removal and the Making of the National Parks*. New York: Oxford University Press, 1999.

Staiti, Paul J. *Samuel F. B. Morse*. Cambridge: Cambridge University Press, 1989.

Standage, Tom. *The Victorian Internet: The Remarkable Story of the Telegraph and the Nineteenth Century's On-line Pioneers*. New York: Walker, 1998.

Stanton, William Ragan. *The Leopard's Spots: Scientific Attitudes Toward Race in America, 1815-59*. Chicago: University of Chicago Press, 1960.

Stephens, Carlene E. *Inventing Standard Time*. Washington, DC: National Museum of American History, Smithsonian Institution, 1983.

——. *On Time: How America Has Learned to Live by the Clock*. Boston: Bullfinch, 2002.

Stevens, Edward W., Jr. *The Grammar of the Machine: Technical Literacy and Early Industrial Expansion in the United States*. New Haven, CT: Yale University Press, 1995.

Stoehr, Taylor. *Hawthorne's Mad Scientists: Pseudoscience and Social Science in Nineteenth-Century Life and Letters*. Hamden, CT: Archon Books, 1978.

Strand, Ginger. *Inventing Niagara: Beauty, Power, and Lies*. New York: Simon and Schuster, 2008.

Stover, John F. *American Railroads*. 2nd ed. Chicago: University of Chicago Press, 1997.

Sturken, Marita, Douglas Thomas, and Sandra J. Ball-Rokeach, eds. *Technological Visions: The Hopes and Fears That Shape New Technologies*. Philadelphia: Temple University Press, 2004.

Sukhdev, Pavan. *The Economics of Ecosystems and Biodiversity: An Interim Report*. Bonn, Ger.: United Nations Environment Programme, 2008.

Sweeney, J. Gray. *Artists of Michigan from the Nineteenth Century: A Sesquicentennial Exhibition Commemorating Michigan Statehood, 1837-1987*. Exh. cat. Muskegon, MI: Muskegon Museum of Art, 1987.

Taylor, Frederick Winslow. *The Principles of Scientific Management*. New York: Harper and Brothers, 1911.

"The Collection and Preservation of Plants." *Godey's Magazine and Lady's Book* 66 (April 1863): 350.

Rosenberg, Nathan. "Technological Interdependence in the American Economy." *Technology and Culture* 20, no. 1 (January 1979): 25-50.

Thomson, Keith. *The Legacy of the Mastodon: The Golden Age of Fossils in America*. New Haven, CT: Yale University Press, 2008.

Thompson, Robert Luther. *Wiring a Continent: The History of the Telegraph Industry in the United States, 1832-1866*. Princeton, NJ: Princeton University Press, 1947.

Thoreau, Henry David. *Walden*, 1854; rpt., New York: Heritage Press, 1939.

Thoreau, Henry David. ed. Walter Harding. *Walden: An Annotated Edition*, Boston: Houghton Mifflin, 1995.

Thornton, Tamara Plakins. "Cultivating the American Character: Horticulture and Moral Reform in Antebellum America," *Orion Nature Quarterly* 1, no. 4 (Spring 1985): 10-19.

——. *Cultivating Gentlemen: The Meaning of Country Life among the Boston Elite, 1785-1860*. New Haven, CT: Yale University Press, 1989.

Trescott, Martha Moore, ed. *Dynamos and Virgins Revisited: Women and Technological Change in History—An Anthology.* Metuchen, NJ: Scarecrow Press, 1979.

Trefethen, James B., *Americans and Their Guns: The National Rifle Association Story through Nearly a Century of Service to the Nation.* Edited by James E. Serven. Harrisburg, PA: Stackpole, 1967.

Trowbridge, John. "Science from the Pulpit." *Popular Science Monthly,* April 1875, 734–39.

Turkle, Sherry, ed. *Falling for Science: Objects in Mind.* Cambridge, MA: MIT Press, 2008.

Twain, Mark. *A Connecticut Yankee in King Arthur's Court.* New York: Harper and Brothers, 1889.

Tyndall, John. "Crystalline and Molecular Forces." *Popular Science Monthly,* January 1875, 257–66.

Tyler, Ron. *American Frontier Life: Early Western Painting and Prints.* New York: Abbeville Press, 1987.

Uselding, Paul. "Elisha K. Root, Forging and the 'American System.' " *Technology and Culture* 15, no. 4 (October 1974): 543–68.

Van Dulken, Stephen. *Inventing the Nineteenth Century: One Hundred Inventions That Shaped the Victorian Age from Aspirin to the Zeppelin.* New York: New York University Press, 2001.

Vlach, John Michael. *Plain Painters: Making Sense of American Folk Art.* Washington, DC: Smithsonian Institution Press, 1988.

Vose, George L. *Manual for Railroad Engineers and Engineering Students.* Boston: Lee and Shepard, 1875.

Wahl, Paul, and Donald R. Toppel. *The Gatling Gun.* New York: Arco, 1965.

Wajcman, Judy. *Feminism Confronts Technology.* University Park: Pennsylvania State University Press, 1991.

Walls, Laura Dassow. *Emerson's Life in Science: The Culture of Truth.* Ithaca, NY: Cornell University Press, 2003.

——. *Seeing New Worlds: Henry David Thoreau and Nineteenth-Century Natural Science.* Madison: University of Wisconsin Press, 1995.

——. *The Passage to Cosmos: Alexander von Humboldt and the Shaping of America.* Chicago: University of Chicago Press, 2009.

Walther, Susan Danly. *The Railroad in the American Landscape, 1850–1950.* Exh. cat. Wellesley, MA: Wellesley College Museum, 1981.

Ward, James A. *Railroads and the Character of America, 1820–1887.* Knoxville: University of Tennessee Press, 1986.

Ware, Norman. *The Industrial Worker, 1840–1860: The Reaction of American Industrial Society to the Advance of the Industrial Revolution.* Chicago: Ivan R. Dee, 1990.

Warner, Deborah Jean. *Perfect in Her Place: Women at Work in Industrial America.* Washington, DC: Smithsonian Institution Press for National Museum of American History, 1981.

——. "Science Education for Women in Antebellum America." *Isis* 69, no. 1 (March 1978): 58–67.

——. "Women Astronomers." *Natural History* 58, no. 5 (May 1979): 12–26.

Webb, William Edward. *Buffalo Land.* Cincinnati and Chicago: E. Hannaford, 1872.

Webber, C. W. *The Hunter-Naturalist; Romance of Sporting or, Wild Scenes and Wild Hunters.* Philadelphia: Lippincott, Gambo, 1852.

Wiebe, Robert H. *Self-Rule: A Cultural History of American Democracy.* Chicago: University of Chicago Press, 1995.

Weisskopf, Victor F. *Knowledge and Wonder: The Natural World as Man Knows It,* 2nd ed. Cambridge, MA: MIT Press, 1979.

Whalen, Matthew D. "Science, the Public, and American Culture: A Preface to the Study of Popular Science." *Journal of American Culture* 4, no. 4 (Winter 1981): 14–26.

Whalen, Matthew, and Mary F. Tobin. "Periodicals and the Popularization of Science in America, 1860–1910." *Journal of American Culture* 3, no. 1 (Spring 1980): 195–203.

Whelan, Richard. *The Book of Rainbows: Art, Literature, Science and Mythology.* New York: First Glance Books, 1997.

White, Frances Emily. "Woman's Place in Nature." *Popular Science Monthly*, January 1875, 292–301.

White, John H., Jr., *American Locomotives: An Engineering History, 1830–1880*. Baltimore: Johns Hopkins University Press, 1968.

Whiting, Cécile. "Trompe l'oeil Painting and the Counterfeit Civil War." *Art Bulletin* 79, no. 2 (June 1997): 251–69.

Whitney, Gordon G. *From Coastal Wilderness to Fruited Plain: A History of Environmental Change in Temperate North America, 1500 to the Present*. Cambridge: Cambridge University Press, 1994.

Wiebe, Robert H. T*he Search for Order, 1877–1920*. New York: Hill and Wang, 1967.

Wilentz, Sean. *The Rise of American Democracy: Democracy Ascendant, 1815–1840*. New York: W. W. Norton, 2007.

Williams, Michael. *Americans and Their Forests: A Historical Geography*. Cambridge: Cambridge University Press, 1989.

Williams, Katherine, et al., *Women on the Verge: The Culture of Neurasthenia in Nineteenth-Century America*. Exh. cat. Stanford, CA: Iris and B. Gerald Cantor Center for Visual Arts, 2004.

Willard, Emma. *Geography for Beginners*. Hartford, CT: Oliver D. Cooke, 1826.

Willard, John Ware. *Simon Willard and His Clocks*. New York: Dover, 1968.

Williams, James C., comp. *At Last Recognition in America: A Reference Handbook of Unknown Black Inventors and Their Contributions to America*. Chicago: B.C.A., 1978.

Wilson, Leonard G., ed. *Benjamin Silliman and His Circle: Studies on the Influence of Benjamin Silliman on Science in America*. New York: Science History, 1979.

Wilson, Stephen. *Art and Science Now*. London: Thames and Hudson, 2010.

Winsor, Mary P. *Reading the Shape of Nature: Comparative Zoology at the Agassiz Museum*. Chicago: University of Chicago Press, 1991.

Wolf, Bryan J. "The Labor of Seeing: Pragmatism, Ideology, and Gender in Winslow Homer's *The Morning Bell*." *Prospects* 17 (1992): 273–318.

Wood, Gordon S. *Empire of Liberty: A History of the Early Republic, 1789–1815*. New York: Oxford University Press, 2009.

Worman, Charles G. *Gunsmoke and Saddle Leather: Firearms in the Nineteenth-Century American West*. Albuquerque: University of New Mexico Press, 2005.

Worster, Donald. *Nature's Economy: A History of Ecological Ideas*. 2nd ed. Cambridge: Cambridge University Press, 1994.

——. *The Wealth of Nature: Environmental History and the Ecological Imagination*. New York: Oxford University Press, 1993.

Wosk, Julie. *Breaking Frame: Technology and the Visual Arts in the Nineteenth Century*. New Brunswick, NJ: Rutgers University Press, 1992.

Wright, Helen. *Sweeper in the Sky: The Life of Maria Mitchell, First Woman Astronomer in America*. Clinton Corners, NY: College Avenue Press, 1997.

Wright, Mabel Osgood. *The Friendship of Nature: A New England Chronicle of Birds and Flowers*. Edited by Daniel J. Philippon. Baltimore: Johns Hopkins University Press, 1999.

Zboray, Ronald J. "Antebellum Reading and the Ironies of Technological Innovation." *American Quarterly* 40, no. 1 (March 1988): 65–82.

Zochert, Donald. "Science and the Common Man in Ante-Bellum America." *Isis* 65, no. 4 (December 1974): 448–73.

✣ Index